T0183470

SpringerBriefs in History of Science and Technology

Series Editors

Gerard Alberts, University of Amsterdam, Amsterdam, The Netherlands

Theodore Arabatzis, University of Athens, Athens, Greece

Bretislav Friedrich, Fritz Haber Institut der Max Planck Gesellschaft, Berlin, Germany

Ulf Hashagen, Deutsches Museum, Munich, Germany

Dieter Hoffmann, Max-Planck-Institute for the History of Science, Berlin, Germany

Simon Mitton, University of Cambridge, Cambridge, UK

David Pantalony, Ingenium - Canada's Museums of Science and Innovation / University of Ottawa, Ottawa, ON, Canada

Matteo Valleriani, Max-Planck-Institute for the History of Science, Berlin, Germany

Tom Archibald, Simon Fraser University Burnaby, Burnaby, Canada

The *SpringerBriefs in the History of Science and Technology* series addresses, in the broadest sense, the history of man's empirical and theoretical understanding of Nature and Technology, and the processes and people involved in acquiring this understanding. The series provides a forum for shorter works that escape the traditional book model. SpringerBriefs are typically between 50 and 125 pages in length (max. ca. 50.000 words); between the limit of a journal review article and a conventional book.

Authored by science and technology historians and scientists across physics, chemistry, biology, medicine, mathematics, astronomy, technology and related disciplines, the volumes will comprise:

1. Accounts of the development of scientific ideas at any pertinent stage in history: from the earliest observations of Babylonian Astronomers, through the abstract and practical advances of Classical Antiquity, the scientific revolution of the Age of Reason, to the fast-moving progress seen in modern R&D;

2. Biographies, full or partial, of key thinkers and science and technology pioneers;

3. Historical documents such as letters, manuscripts, or reports, together with annotation and analysis;

4. Works addressing social aspects of science and technology history (the role of institutes and societies, the interaction of science and politics, historical and political epistemology);

5. Works in the emerging field of computational history.

The series is aimed at a wide audience of academic scientists and historians, but many of the volumes will also appeal to general readers interested in the evolution of scientific ideas, in the relation between science and technology, and in the role technology shaped our world.

All proposals will be considered.

More information about this series at https://link.springer.com/bookseries/10085

Rocco Gaudenzi

Historical Roots of Spontaneous Symmetry Breaking

Steps Towards an Analogy

 Springer

Rocco Gaudenzi
MPI for the History of Science
Berlin, Germany

ISSN 2211-4564   ISSN 2211-4572   (electronic)
SpringerBriefs in History of Science and Technology
ISBN 978-3-030-99894-3   ISBN 978-3-030-99895-0   (eBook)
https://doi.org/10.1007/978-3-030-99895-0

This Springer imprint is published by the registered company Springer Nature Switzerland AG
The registered company address is: Gewerbestrasse 11, 6330 Cham, Switzerland

Preface

This work stems from my fascination for the question of what contributes to a complex discovery and attempts to produce a "cutaway", a core sampling, as it were, of that complexity by way of a conceptual genealogy. Ideally, such a genealogy is a reconstruction of the cultural, material, and cognitive factors that played into the discovery and made it possible. At bottom, it offers a picture of the problems faced by the historical actors, the knowledge resources available to them, and the thinking processes mobilised to address those problems; and it does that by means of an analysis of the selected actors' published and unpublished sources. The knowledge resources are concepts, theoretical techniques, and experimental findings. The thinking processes are here meant as all that explicates in the heuristic strategies adopted when approaching or responding to specific problems—be them the central underlying problems or the secondary and minor ones. In the present story, essential elements of these strategies are mathematical reformulations and generalisations, as well as conceptual transpositions and analogies between distinct realms of physical reality.

The most obvious purpose of our genealogy of spontaneous symmetry breaking is to provide a historical reconstruction which highlights the distinctive aspects of this discovery process. One such aspect is how seemingly distinct (and even "divergent" in the sense that will be specified) branches of knowledge and cultural backgrounds could be mutually fruitful and come to an a priori unexpected confluence. This diving into the turbulent waters of prehistory—and taking the trouble, besides the pleasure, to do so—however, also has the purpose of returning to the surface with a portrait of the logic of scientific discovery (or better, the logic of the *discoverer*); and in particular, the place that analogical reasoning has in such a logic. Two of the "whys" of this engagement with the heuristic processes are to learn as much as possible about what we could designate as the *prospective* logic of the discoverer, and to understand its *durative*, accreting dimension. As our story here shows, the heuristic work often does not proceed by judging the truth or falsity at each step, but takes place on a level where that dichotomy is suspended, and the end justifies the means, as it were. In contrast to the "context of justification", in the "context of discovery" the reasoning proceeds only partially under the pressure and constraint of logical inference.

In the intellectual journey to spontaneous symmetry breaking, this aspect is reflected in the way the leading historical actor, Yoichiro Nambu, employs the analogy. There, analogical reasoning was not a tool of logical inference, but a means to provide different representations, or redescriptions, of the target object that he was each time interested to know or formalise by identifying it with a source object that was recognised as similar in a few respects, while being dissimilar in all the others. The epistemic gain stemming from such identification came from the fact that the "structure" (a property, an attribute, a relation between parts, etc.) of the source domain was already known, or more evident than it was in the target domain. As we will see, in the present case, the quantum vacuum—in the concrete representation that Paul Dirac had given it—was used by Nambu as a master source for the formalisation of nuclear media, the plasma, and then the superconductor, and in turn, the latter two became, at a later stage, the source objects for the redescription of a more complex quantum vacuum. As happens with any act of representation, these redescriptions were not judged as true or false, strictly physical or unphysical, but as appropriate to faithfully render a given aspect of the target object. This appropriateness lay in how much a given redescription at a given moment made visible and instructed the discoverer on how to formulate a hypothesis, and in how much it extracted and made visible what of the target object had remained hitherto hidden or implicit. The reconstruction of these processes and heuristic strategies hopefully conveys the provisional and prospective logic—as opposed to a stable and "retrospective" one—of one who is grappling with something yet to be discovered. The fact, mentioned above, that in this intellectual journey many-body systems and the vacuum of particle physics, enriched at each step, alternately played the role of target and source domain embodies what we refer to as the durative and accreting dimension of this discovery. In this sense, rather than being the result of a one-shot solution, the discovery was a process characterised by an accumulation of experience and exaptation of a number of theoretical tools.

The facts, concepts, and relations which influence and enter more or less directly into the discovery of spontaneous symmetry breaking are several and, at times, demand from the reader some commitment by virtue of their technical character. Considering this, we have constructed the present treatment with various readerships in mind. Ideally, we wrote it so that readings at multiple levels are possible and, we hope, enjoyable for both the scholarly and non-specialist audience. The advice we give to the reader unacquainted with the subject is to treat the technical details like they would treat, in a novel, those detailed descriptions of people and situations which are meant to transmit an atmosphere rather than the details themselves. Even if the details of some notions, their development, and relations to others may remain obscure to some readers, we have made the best efforts to formulate them so that one can capture their gist, and leave the reading with an impression of the conceptual flow in science and a cutaway of the discovery process. The intention of offering various possible readings implied a constant negotiation, and an effort to use the minimum possible of mathematics, to be self-contained, to leave out as much as possible unnecessary technicalities (or confine them to the footnotes) while preserving the core of the concepts and aims that set the wheels of history in motion. The reader will ultimately judge whether we have managed to do that with some success and whether the effort was worthwhile. As a clarification due for the reader who is puzzled by our use of personal pronouns, although it is obvious that literally speaking "we" used throughout is an "I", the intention of the author is not to ascribe an immodest *pluralis maiestatis* to himself, but, quite the contrary, to transmit the conviction that any intellectual effort is ultimately collective in various degrees.

Among the closest people of this "collective", I wish to thank Alexander Blum for having encouraged the ideas behind the project, grounded some of my fancies, and in general for the support given throughout the development of this work. Without his help and competent advice, the level of technical depth, refinement, and cogency of some of the central steps in the development of spontaneous symmetry breaking would hardly have been achieved. I am also grateful to my friend and partner in discussions, Stefano Furlan, whose mental world, brimming with thoughts and original connections, has acted as a source of inspiration and intellectual stimulus. Among the other members of the research group "Historical Epistemology of the Final Theory Program" at the Max Planck Institute for the History of Science, I thank Núria Munõz Garganté, James Fraser, Sebastien Rivat, Giulia Carini, Pablo Ruiz de Olano, and Adrien de Sutter for their presence as friends and discussion partners. The discussions were very helpful and instrumental in the unfolding of my thoughts, as were the discussions with some of the guests of the research group. A thanks go to Ryn Delgado for the great help with the proofreading of the manuscript. I would also like to thank Prof. Giovanni Jona-Lasinio for having shared precious information on the path of discovery discussed here; and Profs. Hajime Inaba, Hiroshi Ezawa, Daisuke Konagaya, and Dr. Hiroto Kono for having shared their knowledge of science in Japan, as well as their willingness to help with the translations from Japanese. The Library Team of the Max Planck Institute for the History of Science has been of utterly invaluable support. Their patience and kindness, together with their efficiency and precision, deserves to be acknowledged. I can say the same about the team of the University of Chicago Library, University of Illinois Archives, and Institute for Advanced Studies.

A part of this work has been carried out under the auspices of the NWO Rubicon Fellowship project ("The genesis of concepts in mesoscopic physics and their role in the reductionist search for a unified theory"), which I was generously granted by the Dutch Science Foundation (NWO). Considering that my formal training and career prior to this work had been in physics, both the Dutch Science Foundation and the Max Planck Institute for the History of Science acted for me as a bridge to a new career and perspective on the sciences, enabled me to construct a background in the history & philosophy of science and, on a personal level, allowed me to fulfil my long-standing desire to do research on the conceptual dynamics and historical epistemological dimension of science. The Max Planck Institute for the History of Science as a supporting institution and platform of knowledge exchange deserves a special mention. The cultural environment at the Institute and its singular blend of expertise have been the yeast of this transition and professional and human growth process.

Berlin, Germany   Rocco Gaudenzi

Contents

Chapter 1
Introduction and a Few Words on the Methodology

... For it is the same thing to apprehend and to recognise as equal
  Parmenides

And how does Professor Nambu go about solving such a
problem?

He answers, laughing, "I think about it all the time"

Each individual is a species

As a Sort of Overture...

In 1961, a couple of articles entitled "A dynamical model of elementary particles based on an analogy with superconductivity", written by Yoichiro Nambu and Giovanni Jona-Lasinio, marked the discovery of what would soon after be called the phenomenon, or mechanism, of "spontaneous symmetry breaking", and which would become a pillar of the nascent Standard Model of elementary particles as well as of the physics of solids. As the title suggests, the key to the discovery was the relation between a phenomenon discovered in solid state physics and the more fundamental realm of particle physics. At the core of the 'dynamical model' proposed by Nambu and Jona-Lasinio lies the suggestive and somewhat uncanny idea that the ground (or "vacuum") state of the universe and its excitations[1] are analogous to those of a many-body medium, in particular a superconductor, and that by means of the latter one can understand the former. The similarity between these two at first disparate physical "objects" indeed consists in that both result from the same underlying mechanism of spontaneous symmetry breaking and show the analogous set of manifestations: their basic components and the interactions between them being described by an equation

---

[1] In the framework of the quantum theory of fields, the individual particles are conceived as excitations of a quantum field and the ground (or vacuum) state is defined as that state where no excitations/particles are present.

© The Author(s), under exclusive license to Springer Nature Switzerland AG 2022
R. Gaudenzi, *Historical Roots of Spontaneous Symmetry Breaking*,
SpringerBriefs in History of Science and Technology,
https://doi.org/10.1007/978-3-030-99895-0_1

with a given symmetry, they both feature a ground state and elementary excitations—the particles we observe—where that symmetry is spontaneously broken (broken, violated or reduced are used as synonymous). The qualification *spontaneous* is key and is there to indicate that the breaking happens by itself; it is exclusively a consequence of the interactions internal to the physical system and not due to forces applied to it from the outside, a process, by contrast, which is called *explicit* symmetry breaking. In this sense, spontaneous (or "self-consistent", as Nambu would alternatively qualify it at first) is therefore to be regarded as opposed to explicit.

Speaking by way of a concrete example, a ferromagnetic material—an ordinary piece of, say, iron or cobalt—can become magnetic if we lower its temperature below a given value (which depends on the property of the material) or if we apply an external magnetic field to it. In the first case, creating the appropriate conditions, the material magnetises in a random direction as a result of the forces internal to it; while in the latter case, it magnetises along the direction of the external field that we fix. In both cases, the material has broken the rotational symmetry that it had originally, i.e., before lowering its temperature or applying the external field. But the essential difference is that in the spontaneous case, all the directions had a priori the same energy and were equally possible. Much like a pencil on its tip can equally fall in any direction and ends up laying in one among them, the magnet has ended up in one such ground state (the state of lowest energy) but could just as well have ended up in any of the others, as there is no preferred direction written in the equation. In the explicit case, the ground state is only one and thus a priori uniquely determined. The peculiarity of the subtler, spontaneous process of symmetry breaking is that the ground state is necessarily associated with excitations which restore the original symmetry, and would allow the material to go from one ground state to the others. Thus, even if each ground state is by itself non-symmetric, when considering the ground state and the spectrum of excitations, the original symmetry remains. For this reason, one can also say that rather than strictly broken, the original symmetry is in fact hidden.

The fact that we have used a relatively mundane object like a magnet as an illustrative example is not casual. As much as it occurs in the superconductors and in the universe at large, this phenomenon transcends them both and rises up to a general principle. Oddly enough, the more ubiquitous, concrete, and thereby potentially more "evident" instances of the principle (e.g., magnetisation and crystallisation) were discovered through the most exotic and more abstract ones. Said otherwise, two comparatively abstract analogous instances of the underlying general principle have pointed at it, and indicated that the same analogy holds between a much larger variety of physical systems, virtually all those systems which undergo a phase transition, to the point that spontaneous symmetry breaking is what characterises a phase transition at its essence. This full recognition of spontaneous symmetry breaking as a general principle and its wide application in solid state physics begin with Nambu's discovery and are given further impulse by the following works of Jeffrey Goldstone and Philip Anderson (Munõz Garganté 2022).

The uncovering and comprehension of the more concrete through the more abstract is certainly one of the fascinating aspects of Nambu's discovery. The reason why that occurred and the sense of concrete and abstract are hereby clarified. The symmetries that are spontaneously broken in the superconductor and in the universe at large are so-named *internal* symmetries, as opposed to spatio-temporal symmetries. In general, a symmetry of a physical system described by a given equation is expressed by the latter remaining unchanged under some transformation: to say that the system has a given spatio-temporal symmetry means then that the equation describing it remains invariant upon the corresponding transformation of the space and/or time coordinates. This essentially reflects the fact that the physical system does not have a preferred "position" in space or in time, or direction. Remaining with the concrete example above, a magnet is, for instance, described by a mathematical *law* which features the single atomic spins of a material coupled to each other. This law remains unchanged upon an arbitrary rotation of all the spins, and it has thus the same symmetry as a sphere. Above a certain temperature, the material is in a non-magnetic state, a *state* which has the same spherical/rotational symmetry of the law. Below that temperature, the magnet spontaneously magnetises in *one* of the possible directions, and in doing so, it ends up in a state that has lost that symmetry. An internal symmetry is a more abstract analogue, and in fact, it is exclusively applicable to mathematical objects such as wavefunctions and quantum fields. Independently of space and/or time coordinates, an internal symmetry is an invariance of the mathematical law under a transformation of the particles' wavefunction. The internal symmetry in question in the specific case of the superconductor is a symmetry under a change of the phase of the wavefunction of the electrons that make it up.[2] A symmetry of the same kind applies to the equations describing the vacuum of the universe and its excitations, the elementary particles. As these symmetries are the abstract analogues of the spatio-temporal symmetries—for instance, the rotational symmetry in a magnet—their spontaneous breaking is an abstract analogue, too.

Given the different degrees of abstraction characterising these orders of phenomena, one might now wonder: why was the mechanism/phenomenon of spontaneous symmetry breaking discovered at work in those relatively abstract cases (first and foremost in superconductors, and then in particle physics) and only then realised in a concrete analogue like, say, the magnet? This question gives us the chance to rewind the progression of this book, from the river mouth to the spring. As we will see in Chap. 4, the core reason is that the discovery of the phenomenon in superconductors hinged on formal-mathematical resources that applied to and were available only for the case of internal (gauge) symmetries, and which were not there for the spatio-temporal symmetries. On the more "human" side, the mobilisation of such theoretical resources was in turn triggered by the sensitive fact that the breaking of an internal symmetry implies the non-conservation of a charge. In the case of the superconductor, that was electric charge, and its apparent non-conservation was

---

[2] In mathematical terms, the global phase φ of the electron wavefunction $e^{i\varphi}\psi$ is the abstract analogue of an angle or a direction, for instance, the direction of magnetisation of a magnet.

unsettling and discomforting so human and theoretical resources were mobilised to resolve it. As we will see in Chap. 3, however, the tools that allowed for the discovery of spontaneous symmetry breaking in superconductors had not been developed within solid-state physics itself—the realm superconductors belonged to—but within particle physics, and in particular quantum electrodynamics, to be then applied to plasmas and nuclei by a particle physicist of a mathematical kind like Nambu. As we will see in Chap. 2, those applications in turn were the product of a specific approach, philosophical premises, and a focus which belonged to a certain tradition of physics in Japan. Within this conceptual Russian doll, we can see the peculiar role that superconductivity had in the process of discovery: it is a many-body phenomenon like, say, ferromagnetism, and at the same time it spontaneously breaks an internal symmetry which is a prerogative of particle physics (there are only a few other cases in solid-state physics where this happens). It is in this sense also that, as we articulate in one of the reflections in Chap. 5, the superconductor could play the role of a scale model of the universe.

Metaphorically speaking, it took a trip from the land of particle physics to the physics of many bodies (solids and nuclei), from there to the province of superconductivity, and from the latter back to particle physics, before the simpler and more concrete instances of spontaneous symmetry breaking could finally be rationalised in light of the general phenomenon. This is the fascinating path of discovery that we set out to unfold here, to the extent that our single and modest angle affords.

The Spirit of Our Reconstruction

Having introduced the concept of spontaneous symmetry breaking, what it is about broadly speaking, its physical significance, and some aspects of the heuristic and conceptual journey which led to it, we wish now to outline our methodological framework.

As we said at the very beginning, the ingenious analogy, drawn by the physicist Yoichiro Nambu, between superconductivity and particle physics can legitimately be regarded as the genesis of the concept of spontaneous symmetry breaking, and as setting a cornerstone of the process that will lead—by way of a few other figures—in the following few years to the Higgs mechanism on one side, and on the other side to the rationalisation of a plethora of phenomena in solid-state physics as instances of spontaneous symmetry breaking. In this sense, that analogy certainly marked the beginning of the *history* of the concept of spontaneous symmetry breaking, that point of time from which the concept is immortalised as such in the written (published) records and begins to float on the surface of history. What about what happened before? What about the *complex* of circumstances underpinning that analogy? This is the problem of history, of official historical records, when one deals with the mental activity of us human individuals. If we leave most of what precedes and prepares a "written" record uninvestigated, we are left with the wrong impression that analogy emerged from nowhere; or, as we more often cloak it, that it was a sudden flicker of genius. The artifice that this situation creates is the analogue in everyday life to a thought that seemingly just occurs, disconnected from the rest. We say it has "come to our minds" to express that we do not know how it has and are not able to retrace its path. While the very cognitive act might indeed be legitimately considered an infinitesimal flicker—in the case of analogy, it is the identification of the similar between two (or more) dissimilars—and cannot be captured with the means of history, the aim of a historical investigation of a discovery is to reconstruct everything around that cognitive act, and in particular what sets it to occur and how that is the culmination of a process.

The "longue-duree" genealogy that we present in this book wishes to do that by following the world-line, as it were, of the central figure behind the analogy, the Japanese physicist Yoichiro Nambu, across about the span of a decade. In doing so, we realise that the analogy is the final step of a long heuristic journey, the *duree* of which is *longue* relative to the acts of drawing the analogy and deriving its final consequence, in this case the understanding of spontaneous symmetry breaking in its generality. As typically happens when framing an historical event in a longue-duree perspective, what appears as a rupture, a sudden opening or a discrete transition—in this case brought about by the cognitive act of analogy—is seen as the conclusion of a process that has an earlier "origin" (without thereby trivialising its novel content).

As far as the work of the central figure of Yoichiro Nambu is concerned, this origin can be located about ten years before the discovery, at the onset of 1950—Nambu's notebook bears witness to this—with his reconsideration, under the light of then new quantum electrodynamics, of another analogy. This is the analogy between the vacuum and the degenerate Fermi gas (a "solid" composed of electrons at absolute zero temperature), which the physicist Paul Dirac had in turn drawn twenty years earlier and had meanwhile been forgotten. The exhuming of Dirac's hypothesis is, in fact, one among the elements which underpin the development of Nambu's thought and, properly transformed, flow into the discovery of spontaneous symmetry breaking. Other important elements are related, as we will see, to earlier work of Sin-Itiro Tomonaga and other Japanese physicists whose peculiar perspectives and approaches to problems would have a significant bearing on the world-line of Nambu. For these reasons, our account of this conceptual adventure to spontaneous symmetry breaking is made to begin in the early 1930s.

By virtue of the role played in the path of the discovery of spontaneous symmetry breaking both by elements of the Japanese physics tradition and by phenomena from different domains of physics, at bottom this book describes the interplay between systems of thought, cultural traditions, and worlds of physical phenomena. More specifically, these are the interplays between the (broadly-intended) Western and the Japanese systems of thought, as well as that between the physics of elementary particles and the physics of "complex" mesoscopic systems (many-body systems from multi-electron atoms, to molecules, to solids). The first interplay followed as a consequence of the engagement of a few young Japanese scientists with the new physics—quantum mechanics and quantum field theory—that had begun to make its way in Europe in the beginning of the 20th century. The second interplay was based on the analogical nexus established between the two worlds of physical systems and phenomena, largely distinct from each other with regard to energy scales, typical sizes, as well as degree of fundamentality (when considered in a reductionist hierarchy). If they are distinct in this sense, these domains can even be said to be in opposition to each other with regard to the experimental and conceptual tools that are employed to explore and understand them, as well as the research questions that animate them. With a touch of overemphasis, on the one hand, the physics of particles is a *high-energy* physics: it aims at *breaking* matter *apart* in the attempt to explore the *elementary* components and interactions between a *few* of them, using high energies to do that. Mesoscopic physics, on the other hand, is the physics which appears at *low energies* and relatively large scales: it aims at the phenomena that emerge from *putting* the elementary constituents *together* into *complex* wholes; and chases thereby the low-energy conditions (low temperatures being one of them) in which the interactions between *many* such constituents best show up.

The divergence of problems, tools, and intents notwithstanding, these two domains of physics did cross-fertilise each other and found a convergence point in spontaneous symmetry breaking. Brought about in its generality both by the concerted action of matter physics (low-energy) and particle physics (high-energy), spontaneous symmetry breaking was a phenomenon found to occur in both. Within these two domains of the universe, that mechanism is equally responsible for an extremely low energy phenomenon like superconductivity—whose characteristic energy scale is about 10^{-8} eV—and for the mass of elementary particles—whose characteristic energy spans from about 10^6 to 10^{10} eV. In that process of mutual fertilisation, analogical reasoning was the heuristic strategy through which the knowledge resources (conceptual schemes and mathematical techniques) developed for open problems and phenomena in the two distinct domains were put to serve one another, and by which some of these phenomena were eventually recognised as instances of the same underlying mechanism of spontaneous symmetry breaking.

Considering this variety of theories, phenomena, problems, tools, scales of space and energy that explicitly or implicitly took part in the development of spontaneous symmetry breaking, that development becomes a machinery to perform a core sampling in the physics of the 20th century: spontaneous symmetry breaking acts somehow as the Rhine river does in the eponymous work of Lucien Febvre (1998). It goes without saying that the histories of those subjects have to a large extent already been dealt with in dedicated accounts. On the front of low-energy physics, these include the developments of superconductivity (Cooper and Feldman 2010; Dahl 1984; Gavroglu 1995), phase transitions and collective phenomena and related mathematical methods (Hiley and Peat 2012; Hoddeson et al. 1992), as well as low-temperature physics (Eckert et al. 1992; Gavroglu and Goudaroulis 1984), to name just a few of the references used here. On the front of high-energy physics, these are the histories of nuclear physics and the approximation methods used therein (Amaldi 2004; Brown et al. 1989; Brown and Rechenberg 1996), quantum electrodynamics (Schweber 2020), and the development of standard model (Brown et al. 1992). This work makes large use of these and other secondary sources, but what distinguishes it from them is the focus on the role that those elements played in the conceptual development of spontaneous symmetry breaking and each of them was instrumental to its discovery. With the purpose of further rendering this interplay, we have also made an effort to discuss at least part of the historical motives and triggers at the origin of the theories, phenomena, speculations, problems that then took part in the development of spontaneous symmetry breaking. Such second-order influences, as it were, include the peculiar tradition of modern physics emerging in Japan, reconstructed and discussed in Chaps. 2 and 5, as well as how some advancement in physics enabled new technologies (e.g., cryogenics, cosmic rays and accelerators) which in turn enabled new physics in a positive feedback cycle.

Ours is not the first reconstruction of the development of spontaneous symmetry breaking, and to understand how the present work contrasts with and complements the other reconstructions we would like to briefly review them here. The first historical account of the discovery of spontaneous symmetry breaking is the one offered by Brown and Cao (1991). This discusses in some detail especially the last part of the heuristic development from Nambu's perspective, and has constituted a useful starting point for the present treatment. However, the work does not discuss how Nambu's background and previous work were instrumental to that development, nor the key role that elements peculiar to the schools of Tomonaga, and Yukawa and Sakata played in it. To some extent complementary to that work is Laurie Brown's article "Yoichiro Nambu: the first forty years" (Brown 1986). As the title suggests, the article offers an overview of the personal and professional aspects of the physicist's life in the period of interest (roughly from the 1920s to early 1960s). While a starting point to reconstruct Nambu's attitude towards the physical problems, as well as some of the preliminaries to spontaneous symmetry breaking, the article is an overview on the individual and, in fact, does not focus in particular on spontaneous symmetry breaking. A more recent and broader historical account on symmetry breaking, from Euler to the Higgs boson, is given by Sardella (2012). This covers a large part of the themes and steps that are propaedeutical to spontaneous symmetry breaking and has been a very valuable and informative source for this work. Its breadth and framing, however, allow even less than the aforementioned studies a focus on Nambu's longue-duree heuristic pathway and the role of analogy. More recently, partly triggered by the experimental discovery of the Higgs boson, a few studies reconstruct the development of the Higgs mechanism (Borrelli 2015; Karaca 2013). In doing so, they chart the central role that spontaneous symmetry breaking had in it and, in general, in the Standard Model of elementary particles. These reconstructions roughly start where ours ends, and are to be considered in this sense complementary to it.

Building on these studies, here we have decided to frame the development of spontaneous symmetry breaking by combining the dimension of a few individual heuristic pathways with the broader (we could say collective) conceptual dimension forming their background. By making use of prolepses, analepses, and digressions, we have made the effort to organise the historical flow around a single narrative line largely made up of a stretch of the "world-lines" of Tomonaga first and then Nambu—a "soggettiva" on them in cinematic language. We have done that by putting into correspondence the contents of a number of scientific publications and informal—and largely unknown—communications in the Japanese "tabloid" *Soryushiron Kenkyu* with the archival material, the material collected at the Japan-USA collaborative workshop (chronicles, recollections, round table discussions) (Brown et al. 1991) as well as various Oral Histories Interviews of the American Institute of Physics (Ashrafi 2004; Hoddeson 1999; Kojevnikov 1999; Warnow and Williams 1974). As far as the archival material, we have taken advantage of the Yoichiro Nambu Papers recently made available for consultation at the University of Chicago Library, as well as the material from other archives (the Archives of the Institute for Advanced Studies in Princeton and the University of Illinois Archives). We hope we were able in this way to portray the genesis of a complex scientific concept and highlight the creative inputs and cultural elements involved in a scientific endeavour.

References

Amaldi, U. (2004). Nuclear physics from the nineteen thirties to the present day. In *Enrico Fermi: The life and legacy* (pp. 151–176). Springer.

Ashrafi, B. (2004). Interview of Yoichiro Nambu by Babak Ashrafi on 2004 July 16. www.aip.org/history-programs/niels-bohr-library/oral-histories/30538.

Borrelli, A. (2015). The story of the Higgs boson: The origin of mass in early particle physics. *The European Physical Journal H, 40*(1), 1–52.

Brown, L., Dresden, M., & Hoddeson, L. (1989). *Pions to quarks: Particle physics in the 1950s*. Cambridge: Cambridge University Press.

Brown, L. M. (1986). Yoichiro Nambu: The first forty years. *Progress of Theoretical Physics Supplement, 86*, 1–11.

Brown, L. M., Brout, H. R., Cao, T. Y., Higgs, P. W., Nambu, Y., Hoddeson, L., Brown, L., Riordan, M., & Dresden, M. (1992). Panel session: Spontaneous breaking of symmetry. In *The rise of the standard model: Particle physics in the 1960s and 1970s* (pp. 478–522). Cambridge University Press.

Brown, L. M., & Cao, T. Y. (1991). Spontaneous breakdown of symmetry: Its rediscovery and integration into quantum field theory. *Historical Studies in the Physical and Biological Sciences, 21*(2), 211–235.

Brown, L. M., Kawabe, R., Konuma, M., & Maki, Z. (1991). Elementary particle theory in Japan, 1930–1960; Proceedings of the Japan-USA collaborative workshops. *Progress of Theoretical Physics, Supplement, 105*.

Brown, L. M., & Rechenberg, H. (1996). *The origin of the concept of nuclear forces*. CRC Press.

Cooper, L. N., & Feldman, D. (2010). *BCS: 50 years*. World scientific.

Dahl, P. F. (1984). Kamerlingh Onnes and the discovery of superconductivity: The Leyden years, 1911–1914. *Historical Studies in the Physical Sciences, 15*(1), 1–37.

Eckert, M., Schubert, H., & Torkar, G. (1992). The roots of solid-state physics before quantum mechanics. In *Out of the Crystal Maze: Chapters from the history of solid state physics*. Oxford University Press.

Febvre, L., Schöttler, P., & Galeotti, A. (1998). *Il Reno. Storia, miti, realtà*. Saggi. Storia e scienze sociali. Donzelli.

Gavroglu, K. (1995). *Fritz London: A scientific biography*. Cambridge University Press.

Gavroglu, K., & Goudaroulis, Y. (1984). Some methodological and historical considerations in low temperature physics: The case of superconductivity 1911–57. *Annals of Science, 41*(2), 135–149.

Hiley, B., & Peat, F. D. (2012). *Quantum implications: Essays in honour of David Bohm*. Routledge.

Hoddeson, L. (1999). Interview of David Bohm by Lillian Hoddeson on 1981 May 8. www.aip.org/history-programs/niels-bohr-library/oral-histories/4513.

Hoddeson, L., Schubert, H., Heims, S., & Baym, G. (1992). Collective phenomena. In *Out of the Crystal Maze: Chapters from the history of solid state physics*. Oxford University Press.

Karaca, K. (2013). The construction of the Higgs mechanism and the emergence of the electroweak theory. *Studies in History and Philosophy of Science Part B: Studies in History and Philosophy of Modern Physics, 44*(1), 1–16.

Kojevnikov, A. (1999). Interview of Philip W. Anderson by Alexei Kojevnikov on 1999 November 23. www.aip.org/history-programs/niels-bohr-library/oral-histories/23362-3.

Munõz Garganté, N. (2022). Ph.D. thesis (in preparation), Technical University Berlin.

Sardella, I. (2012). *Storia della rottura di simmetria: dalla colonna di Eulero al Bosone di Higgs, il lungo cammino di un'idea*. Aracne editrice.

Schweber, S. (2020). *QED and the men who made it: Dyson, Feynman, Schwinger, and Tomonaga*. Princeton Series in Physics. Princeton University Press.

Warnow, J., & Williams, R. (1974). Interview of J. Robert Schrieffer by Joan Warnow and Robert Williams on 1974 September 26. www.aip.org/history-programs/niels-bohr-library/oral-histories/4864-1.

Chapter 2
Traits of an Emerging Tradition: Modern Physics in Japan

In this chapter we witness Tomonaga's two seminal attempts at a description of nuclear interactions, and the developments that follow from it. Both planted during his sojourn in Heisenberg's Leipzig and based on the reinterpretation of methods used in atomic or matter physics, these seeds will respectively grow into Tomonaga's interpretation of renormalisation theory and into his pioneering "collective" approach to many-body problems. These elements, along with Tomonaga's approach to the physical problems, constitute the tradition which Nambu will subsequently build on and take to its most fruitful consequences with the discovery of spontaneous symmetry breaking.

By way of introduction we should specify what we mean here by a "tradition". The notion is in general problematic, and even more so when one practically attempts to identify the supposed beginning of a specific tradition. What we call "tradition" here is what acted as one for those who followed: the complex of thought styles, guiding ideas, approaches to problems, and background philosophies established or conveyed by the figures that constitute the reference for a new generation, either through direct contact (teacher-pupil relationship, collaboration, etc.) or more indirect contact (listening to discussions and seminars, i.e., the oral tradition). In our case, the tradition in question is constituted by the figures of Sin-Itiro Tomonaga, Hideki Yukawa and, partly, Shoichi Sakata and Mitsuo Taketani, a few years junior to the first two. The generation that built on the thinking and results of these figures and was influenced by them is represented by those young physicists, born in the 1920s, who ventured—in fact, in the relatively conservative Japan, dared to—into nuclear and particle physics' fundamental research. Among these young physicists we find Yoichiro Nambu, who is the central protagonist of the discovery of spontaneous symmetry breaking and whose development we will follow in the next chapter. The reconstruction of the traits of this tradition is relevant for our story in that it shows how those views and open problems that are posed by the European-born modern physics and will

© The Author(s), under exclusive license to Springer Nature Switzerland AG 2022
R. Gaudenzi, *Historical Roots of Spontaneous Symmetry Breaking*,
SpringerBriefs in History of Science and Technology,
https://doi.org/10.1007/978-3-030-99895-0_2

enter directly or indirectly the subsequent development of spontaneous symmetry breaking are partly transformed by the Japanese physicists while being assimilated. This "active" assimilation manifests itself in the peculiar course that the views and questions take in Japan, and the original approaches to them adopted by the handful of aforementioned physicists.

The challenge that this reconstruction poses from a historiographical standpoint is not small, since the elements that are propaedeutical to spontaneous symmetry breaking, and will be part of its development, are connected to virtually all the major problems and corners of theoretical physics between the 1930s and the 1950s: from the issues such as those posed by the construction of a relativistic version of quantum theory, to the problem of treating with such a framework physical systems (from nuclei to solids) that feature many bodies and/or a strong interaction along with the complex phenomenology that results.

Considering the breadth, we certainly cannot, and do not intend to, give a systematic exposition of these themes. As far as this chapter is concerned, we rather choose to recount them through the eyes of Sin-Itiro Tomonaga, for how they entered his "field of view" and were approached by him. The choice of Tomonaga, rather than for instance Yukawa, is not casual. Across the course of his career, Tomonaga tackled a range of problems and themes in ways that closely informed Nambu's future development and his approach to the physical problems. This chapter aims to be, in this way, at once an overview of the conceptual threads underlying spontaneous symmetry breaking and a reconstruction of a central part of the tradition Nambu draws from. In doing so, we leave the discussion of the subtler and less technical counterpart of the tradition—incarnated by Yukawa, Sakata, and Taketani—and its influence on the development of spontaneous symmetry breaking to the last chapter.

2.1 Tomonaga's Approach to the Open Questions of Particle Physics

If we had to find a beginning point—with the arbitrariness inherent to all punctual beginnings—to the adventure of thought we are about to tell, we may locate it in the cycle of lectures held in Kyoto and Tokyo by the young Paul Dirac and Werner Heisenberg and promoted by Yoshio Nishina (Konagaya 2020). The year is 1929, Nishina is just back to Japan from an eight-year period spent visiting European universities and laboratories and a durable collaboration with Niels Bohr and colleagues in Copenhagen (Ito 2002; Kim 2007). The idea behind the series of lectures is to introduce the young Japanese students to the principles of the "new" quantum mechanics (non-relativistic and relativistic) and some of its applications through the mouths of two of its major proponents.[1] While it is certainly not the first contact

---

[1] As reported by the Japanese historian of science Daisuke Konagaya (2020), the topic of the lectures are allocated day-by-day: the indeterminacy relations and the physical principles of the quantum theory (Heisenberg, one day), the principle of superposition and the two-dimensional harmonic

between European and Japanese physicists, even less the inception of a more systematic cultural exchange between Europe and Japan—which could be placed at the beginning of the Meiji Era, around 1869—this event nonetheless has a material and symbolic importance, in general and for our story. If we wish, it marks the wider and more "institutional" introduction in Japan of the brand new quantum mechanics, the exposition of the forefront of theoretical understanding of the atomic world, its successes, the open questions and problems. Not that the latter are by then completely unknown and hitherto not addressed in Japan, but the new quantum mechanics is not yet formally taught; it is just cultivated by the relatively few who, like Nishina, have learned one of the European languages, and learned and collaborated directly with the proponents. There are no textbooks on this topic in Japanese, almost no vocabulary to speak about it, and very few people who could teach it at all in an academic ground where classical physics is still the rule (Darrigol 1988; Kim 2007; Ashrafi 2004). Venturing out in that territory is left to the self-study and personal initiative of a handful of the ambitious (Nakane 2019).[2]

In this context, the presence of Heisenberg and Dirac gives a significant impetus to the efforts of linguistic and conceptual "translation" of modern physics in Japan, and coincides with the beginning of a wider participation of Japanese physicists in a field that had hitherto been almost exclusively a prerogative of the Western world. Two of these future participants and most significant contributors are just about to start their endeavour into the discipline: in the audience, awed and puzzled in equal measure, Tomonaga and Yukawa sit as young graduate students, and the role that those lectures played in pushing them, among others, to engage with the frontier problems of the physics of time is not secondary (Brown et al. 1991).

In the years that follow this event, Yukawa will transcribe, study and translate Dirac's and Heisenberg's papers.[3] He will work quite intensely and obsessively, although ultimately unsuccessfully, on the problem of divergences unsettling the nascent quantum theory of fields. As we will see below, in the face of Ockam's razor—*entia non sunt multiplicanda praeter necessitatem*—in this period Yukawa also postulates the existence of a new particle as a carrier of the nuclear force, the π-meson or Yukawa meson that will later make him famous worldwide. As for Tomonaga, he will join the Institute for Physical-Chemical Research (*Riken*, Tokyo) alongside Yoshio Nishina. There he begins to measure himself against various problems in the framework of Dirac's theory of electrons and nuclear physics, at times improving on or extending existing methods and results, at times independently obtaining similar results (see, e.g., Tomonaga's early works in Tomonaga and Miyajima 1976a). Tomonaga and a few others, in the summer of 1935 and under

---

oscillator (Dirac, one day), retarded potential in the quantum theory (Heisenberg, two days), the quantum mechanics of many-electron systems (Dirac, two days), relativity theory of the electron (Dirac, two days), the theory of ferromagnetism (Heisenberg, two days).

[2] As we will see in the next Chapter, this attitude continued to characterise places such as the Tokyo Imperial University also in the later years as, among others, Nambu (1991) and Kinoshita (1991) recollected.

[3] This was communicated to us by Daisuke Konagaya on the occasion of a seminar held at the Max Planck Institute for the History of Science in March 2021.

the direction of Nishina, will also accomplish the first translation into Japanese of Dirac's textbook "The principles of quantum mechanics". By way of that translation, the small group effectively constructs a Japanese terminology to speak about and visualise quantum phenomena, which has continued to evolve to this day.[4] A very exacting task, as Tomonaga later recollected (Tomonaga 1968), this conceptual and linguistic work is carried out by the small group for the sake of the "many [future] physics' students of our country".

As had happened earlier with Western-born philosophy in Japan,[5] the seed of European physics planted in another fertile ground would grow into other fruits. We will see here how that growth unfolded by following the subsequent path of Tomonaga for how it intertwines with the path of Yukawa and Western physics.

2.1.1 At Heisenberg's and the Renewed Interest in Nuclear Meson Hypothesis

It is with the background that we have sketched, along with a dose of excitement for his first trip abroad, that Tomonaga arrives in Leipzig in June 1937 with the purpose of spending a number of years with Heisenberg. As noted in his *German diary*, the discussions in Heisenberg's group concerned vacuum polarisation, but also had, shortly before his arrival, started gravitating around recent cosmic rays experiments, looking at them through the hypothesis of Yukawa's meson and its decay (Brown et al. 1991). As we have mentioned, in 1935 Yukawa had boldly hypothesised that nuclear forces are transmitted by the exchange of a massive particle, a massive analogue, Yukawa had reasoned, of the photon that carried the electromagnetic force. The idea of invoking such a particle was seductive, but seemed a preposterous move, an unnecessary multiplication of the known elementary entities—by then, just electrons, protons, neutrons, neutrinos, and photons. In fact, such a particle had not been observed and the hypothesis had thus not been given much credit in the scientific community until two experiments by Carl Anderson and Seth Neddermeyer were published in August 1936 and May 1937. The experiment reported unusual tracks of "penetrating and [...] heavily ionizing particles that usually arise from a type of nuclear disintegration not heretofore observed" (Anderson and Neddermeyer 1936) and decayed into an electron and an anti-neutrino. Heisenberg and Hans Euler had at that point suggested extracting the lifetime of the ionizing particle assuming these were Yukawa's meson. A plausible agreement was found between the value extracted from the experiments and the value calculated by Yukawa and Sakata (1939) on the basis of the nuclear

---

[4] The physicist Hiroshi Ezawa has tracked the evolution of about thirty terms across five different editions of Dirac's Principles starting from the first (private communication with Hiroshi Ezawa).

[5] In an essentially analogous task but in the realm of philosophy, Tomonaga had been preceded by his father Sanjuro, who had compiled in 1905 the first dictionary of Western philosophy in Japan. Sanjuro Tomonaga, along with Nishida Kitaro and Tanabe Hajime, founded the Kyoto School of Philosophy, an original combination of Western philosophy and Buddhist thought.

coupling and the beta decay rate. This agreement scored an important goal in *favour* of Yukawa's theory and encouraged many physicists to identify those penetrating particles with Yukawa's meson.

As the subsequent theoretical papers reveal, the interpretation of the cosmic-ray experiments—the only available particle accelerators of the time—is, however, far from a trivial matter: lumping together the effects of at least two interactions, the long-range Coulomb and the short-range nuclear interaction, the numerical results obtained depend to a large extent on the models assumed for the interaction, as well as the theories that are used in the data analysis. Theory and experiments are, as a result, in a constant feedback loop where theoretical improvements change the way in which the same data are analysed. Such step-by-step theoretical improvements allowed, on the one hand, for confirming the "penetrating particles" to be heavier than electrons— against competing explanations—but also revealed a meson-nucleon scattering rate at high energy much smaller than theoretically expected. With a nuclear coupling estimated according to the Yukawa model, these particles should not have been as penetrating. An argument thus arises *against* the identification.

During the course of the year 1938, corresponding with Tomonaga's second year in Leipzig, some of the attention shifts towards this inconsistency, which becomes one of the major reasons to hold reservations against the meson theory. There is, how-ever, a caveat, that Heisenberg, among others, signals: the point is that the scattering cross section in question is calculated, in absence of an alternative approximation technique, with customary perturbation theory (Maki 1991). This presupposes a small meson-nucleon coupling, and in practice automatically allows for considera-tion only of the processes where zero, one or two mesons are involved. Because the nuclear coupling is in fact *not* small—and, correspondingly, the processes involv-ing *many* mesons are actually the dominant ones[6]—the perturbative treatment also greatly underestimates the self-field around the nucleon, and consequently neglects the possible influence of this self-field on the meson-nucleon scattering. This leaves open the possibility that the self-field might be responsible for the lowering of the meson-nucleon scattering. Firmly convinced of the need to go beyond the pertur-bation method hitherto used, Heisenberg drafts a (classical) mechanical toy model where the mesons of the self-field are strongly coupled with the nucleon's spin (and angular momentum) (Heisenberg 1939). This gives the large *inertia* to the nucleon which suppresses the scattering of an incoming high-energy meson, and yields a value closer to the experiments.

---

[6] For the reader that is unfamiliar with these matters, the reason why the strength g of a force is correlated with the number of virtual particles which mediate that force can be intuitively understood by noticing that, in quantum field theory, the probability of a process featuring n virtual particles is roughly proportional to g^n. So, if g is small ($g \ll 1$)—as in, e.g., electromagnetism—the processes with a low number of interactions (appearing in the first orders in perturbation theory) will occur more often than those featuring several interactions (higher orders in perturbation theory); on the contrary, if g is large ($g \gg 1$)—as in the case of nuclear force—the opposite happens. Hence, in general, the more intense the force, the more probable are the processes involving a high number of virtual particles. This calls for a (non-perturbative) approximation method that deals with many mesons at once, and, as we will see, motivates the analogy that Tomonaga draws with many-body problems.

Exhorted by Heisenberg's view on the crucial importance of the field reaction to properly reproduce nuclear processes, and the necessity of an alternative to the perturbative approximation that could properly treat the processes involving many mesons, Tomonaga sets out to derive a closed expression for the (divergent) self-energy of the nucleon, with the idea of then subtracting it in the calculation of the scattering processes (Maki 1991). In order to do that, Tomonaga starts by constructing a quantum mechanical version of the model proposed by Heisenberg.

The situation in Leipzig in those days of 1939 is still seemingly quiet. Without much clamour, Hitler's army trails the city, in its march towards the Sudeti mountain range. Shortly after, Tomonaga sits on a train to Hamburg, with his request for a one-year extension denied, and World War II breaking out.

2.1.2 A Creative Application of the Hartree Method

Back in Japan after a few-month journey across the two oceans, leaving behind a Europe on fire, Tomonaga learns about a paper, just issued by Gregor Wentzel, featuring a successful non-perturbative technique to address nuclear scattering. Using a canonical transformation similar to that used previously by Pauli and Fierz, which "dressed" the electron with the cloud of virtual (low-energy) photons, Wentzel had transformed the original Hamiltonian for the meson-nucleon system and expanded it in a power series of the *inverse* meson-nucleon coupling constant ($\frac{1}{g}$) (Wentzel 1940). The resulting series had the desired characteristic of being rapidly convergent for strong coupling (i.e., $g \gg 1$), with a first term, the new unperturbed term, no longer representing the bare nucleon, but the nucleon *together with* its inseparable cloud of mesons in an arbitrary number. This new physical (or dressed) nucleon was shown to have its own stable excitation spectrum and could therefore, like the dressed electron of Pauli and Fierz, effectively be treated as a standard particle. While in fact suppressing the scattering at high-energy, Wentzel's method still did not match the experimental trend.

Assuming Wentzel's concept of the physical nucleon to be valid, but considering the particle aspect of the mesons rather than the more customary wave-like one, Tomonaga travels an alternative path recognising that the problem is "a *kind of* many body problem" which "will be solved by means of the Hartree method." (Tomonaga 1947). More specifically, he proposes an approximation method which, in his own words (Tomonaga 1955b)[7]:

> [...] consists in finding the most accurate wave-function of the mesotron [i.e., meson] in the form of a product and can be compared with the well-known Hartree method.

---

[7] The article, first of a pair, was published in 1941 in German as a report on the "local" *Riken* Journal (Scientific Papers of the Institute for Physical-Chemical Research) and can be found in Tomonaga and Miyajima (1976a). Here we refer to the republication (in 1955) on the Supplement of Progress of Theoretical Physics, which is easier to find. The translation from German is ours.

In doing so, he takes into account that (the emphasis is ours):

> Since [the Hartree method] has been applied only to problems concerning atoms and molecules and never to the field theory, it is desirable first to bring the latter theory into a form in which closer *analogy* with the former is obtained.

This operation is carried out by Tomonaga essentially as follows. To begin with, the nucleon Φ is represented as a set of wavefunctions, where each single wavefunction represents the state of zero mesons, one meson, two mesons, and so on, around the nucleon, and where each meson has its own momentum and charge.[8] By viewing these *virtual* mesons surrounding the bare nucleon as analogous to the swarm of *real* electrons dispersed around the core of an atom, each of these wavefunctions is then written out—transposing the simplification adopted by Hartree in the atomic case (Hartree 1928)—as the *product* of simple single-meson wavefunctions, uncorrelated with one another. As the number of mesons is not fixed, in contrast to the number of electrons in the Hartree method, the virtual character of the mesons is the disanalogical aspect. What allows Tomonaga to draw the analogy is then the inclusion of the number of mesons among the arguments of Φ. Expressing the expectation value of the meson-nucleon Hamiltonian for a nucleon wave-function of this form and minimising it as done in the Hartree method, one obtains the most probable configuration (in terms of a number of mesons, their momentum and charge) and its energy expressed as a function of the nuclear coupling strength. Since it neglects the correlations between the mesons, the energy obtained in this way is the sum of the *average* energies of the single mesons all in the *same* nuclear potential, and represents the bulk of the self-energy of the physical nucleon.

Adopting from Hartree the ingenious simplifying hypotheses of his model and adapting it to virtual particles as described, Tomonaga effectively provides a field theoretical Hartree-Fock reformulation of the nucleon's self-energy, satisfying in this way the original desire for a more concrete (particle) picture of the physical nucleon, and at the same time circumventing the limitations of perturbation theory with an intrinsically many-body framework. The validity and generality of such reformulation, beyond its intuitive qualities, are ultimately confirmed by the fact that in both the strong ($g \gg 1$) and the weak coupling limit ($g \ll 1$), Wentzel's and the self-energy expressions obtained through perturbation methods are respectively recovered. For this reason, this model comes to be called "intermediate coupling theory".

The essential traits of the Hartree-Fock model that we have depicted are first laid out in a couple of conferences in Tokyo, to be then systematically written up in a paper published in the *Riken* journal in 1941 under the title "Zür Theorie des Mesotrons. I" ["On the theory of mesotrons, I"] (Tomonaga 1955b; Tomonaga and Miyajima 1976a). The first of a series of instalments, the model will undergo a series of extensions, corrections, and adaptations over the following decade, in an attempt at fulfilling the promise of providing that functionally accurate meson-nucleon scattering cross section which had stimulated it in the first place. We will follow part of

---

[8] For a concise explanation see Maki (1991).

this development, focusing in particular on how it ties with the development of quantum electrodynamics in the peculiar course that the latter followed in the Japanese context under Tomonaga's sway. This will provide the first fundamental element of intersection and commonality between the domain of nuclear phenomena and that of the practices of renormalisation that would be developed for electromagnetic phenomena; and will illuminate subtler aspects of the thinking style of Tomonaga that will reveal themselves as an important ingredient in some members of his school, including Nambu.

2.1.3   The Tomonaga-Nishina Seminars and the Development of Quantum Field Theory

Tomonaga's European sojourn was intended as a temporary stay, so, once back in Japan, he rejoined the former group at the Institute for Physical and Chemical Research (*Riken*) in Tokyo, returning to work under the direction of his elective academic father Yoshio Nishina (Brown et al. 1991). As we mentioned at the beginning, Nishina was an important figure for physics in Japan who had an active exchange with physicists such as Bohr, Heisenberg, and Dirac after his long immersion in the "quantum-mechanical Europe". It was Nishina himself who had indeed arranged the stay of his protégé Tomonaga in Leipzig. As he had come back to Japan from Europe, in 1929 Nishina had established one of the *Riken*'s laboratories: his was part of a constellation of independent laboratories each of which was directed by a chief scientist who was given considerable autonomy to manage research topics, personnel and budget.[9] This so-called "laboratory system" allowed for human and scientific interactions among scientists from the different laboratories, and promoted them, for example, through a trans-disciplinary joint colloquium. Among Tomonaga's most vivid reminiscences of the colloquia were interactions with physicists working on semiconductors, nuclear spin relaxation, and approximation methods in solid state physics. Within the constellation of laboratories, Nishina's covered a variety of topics around atomic physics: they were performing experimental research on cyclotrons, radio-isotopes, and ionization chambers & cosmic rays, fundamentally as well as in their biological and medical uses (Kanamori 2016). Starting from 1932, the Nishina's laboratory had an annexed theoretical physics division formed around Tomonaga, which featured Sakata, Kobayashi (later the founder of the first theoretical solid-state physics group in Kyoto (Tomonaga 1968)), and a few others (Brown et al. 1991). In this "scientists' free paradise", informal and active discussions took place daily, modelled after the European groups that had been visited by the first Japanese physicists abroad, who admired the—exotic in their eye—"Kopenhagener Geist" (Ito 2002).

---

[9] Some historical information on the institutional structure of *Riken* are on the Institute's web page https://www.riken.jp/en/about/history/story/ [last accessed on April 22, 2022]. A review of its structure and activities can be found in Kanamori (2016).

Upon his return from Leipzig, in early 1940, Tomonaga finds the same environment, but now with extended ranks: younger collaborators like Satoshi Watanabe, and theorists from Yukawa's group like Mitsuo Taketani are part of the group, but also joined by the most wilful among the fresh graduates from the Tokyo Imperial University like Tatsuori Miyazima. In this new constellation, with Tomonaga taking the lead of the traditional weekly seminars of the Nishina Laboratory, *Riken* becomes one of only two places in Japan where research on the meson theory of Yukawa is systematically pursued, both experimentally and theoretically (Miyajima 1976a). The seminars, which start to be known as "Tomonaga-Nishina seminars", are now no longer reserved only for the internal members of the theory group, but are also, in keeping with the principle of openness and participation, formally opened to the students of the surrounding universities who are most curious to venture into the "exotic" premises of high-energy nuclear physics. In fact, as a result of a natural selection, only a dozen dare to participate: among them, but sitting and listening hidden at the back of the room, one could see, from time to time (Ashrafi 2004; Nambu 1991).

In continuity with the work initiated in Germany, at the seminars in this period 1940–41 Tomonaga discusses with the experimentalists the increasingly accurate results of the meson-nucleon scattering cross section (Brown et al. 1991), the problematic discrepancy between them and the cross section calculated with perturbative methods, and the alternative Hartree-Fock model of the nucleon aimed at accounting for this discrepancy by including self-energy effects non-perturbatively (see previous Section). That discrepancy is, however, not the only difficulty of the meson theory now that another theoretical inconsistency undermines precisely the result that had motivated the diffused enthusiasm following the discovery of the cosmic ray meson. Where indeed the lifetime of the meson calculated then had encouragingly matched the one estimated experimentally, revised calculations—independently performed by Yukawa and collaborators, as well as by British theoreticians in 1940—reveal a difference of two orders of magnitude: a significant mismatch, hardly ascribable to experimental uncertainty, which made the identification of the observed particle with the meson of Yukawa unjustified and the Yukawa hypothesis itself lose ground (Tomonaga 1955a; Brown et al. 1991).

With the intent of discussing such matters altogether, while the Pacific front of World War II breaks out with the Japanese offence to Pearl Harbour and Japan grows increasingly distant from the international scientific community, the seminars at *Riken* are extended, after a proposal by Nishina, Tomonaga and Taketani, into gatherings of the so-called "Meson Meetings" (Proceedings of the Japan-USA Collaborative workshop 1991). These are a series of inter-university symposia to be held regularly every four or five months (most of them at *Riken*) together with physicists from the Kyoto and Osaka universities. In spite of the name and the intent, at least some of these meetings should be interpreted by way of a synecdoche (see the list of lectures in Proceedings of the Japan-USA Collaborative workshop 1991): through meson theory, what is discussed is the very basis of an elementary particle theory, how the future quantum field theory should look, and whether the paradigm itself needs to be entirely revised. In that sense, the discussion gravitates around the main issue affecting the quantum theory of fields: the problem of the infinite field reaction / self-energy. On that ground, Tomonaga and Yukawa, the two souls of Japanese physics and former schoolmates, lead the debate along two methodologically different pathways. Tomonaga aims at a concrete picture of the dynamics of the self-field and is founded on a critical analysis of the approximation *methods*. Yukawa, on the other hand, evidently starts from a more fundamental questioning of the basic *concepts* of elementary particle and field which, according to him, are in irreparable conflict (Maki 1991; Brown et al. 1991).[10] A brief focus on two of the meetings will be instrumental to frame these two philosophies, and give a glimpse of some important concepts and modes of operation adopted in reaction to the impasse.

In the first such meeting, in June'41, Tomonaga presents the core idea of the Hartree-Fock non-perturbative method and its first concrete application to obtain a closed formula for the scattering cross section which can be compared with experiments. In the formula, obtained in the idealised case of an infinitely heavy nucleon and under a couple of other additional simplifying assumptions, the self-field of the nucleon explicitly appears in the denominator of the expression, confirming its role in the suppression of the cross-section, as originally intuited by Heisenberg (Tomonaga 1976a). While encouraging in some respects, the scattering cross sections are still staked on a cut-off procedure, undesirable for its arbitrariness as well as its lack of relativistic invariance. This fact makes his non-perturbative approach appear no closer to a resolution of the problems of infinities than the solutions that had been hitherto proposed in the context of electrodynamics. Tomonaga makes no mystery of this fact as it is well-evident from these words (written in a further communication co-authored with Miyajima in 1943 (Miyajima and Tomonaga 1976):

> In the present stage of relativistic quantum mechanics [...] the well-known divergence difficulties prevent us from treating such problems in a consistent way beyond the scope of the perturbation theory [where the self-energy is effectively neglected]. The field equations of the present theory must ultimately be regarded, strictly speaking, as having no finite solutions at all.

On the other hand, as he reveals shortly after, the reason for his perseverance on the established track of the non-perturbative Hartree-Fock model is motivated as follows:

> It may be, nevertheless, expected that, if we interpret the present theory correctly, it will be a good approximation of the forthcoming theory, and for each solution of the fundamental equation of the latter theory the corresponding solution in the present theory will exist in some way. If we accept this idea the task will be to treat the problem exactly in this sense according to the present theory and compare the results with the experiments.

Beyond the immediate relevance for accounting for nuclear phenomena, the hope is therefore that the meson theory, that had stimulated the adoption of a non-perturbative *approximation* method, could yield a satisfactory solution, or at least a

---

[10] To be more specific, one could say that Tomonaga's strategic attitude to the meson theory (as well as elementary particle theory) was somewhat different from that of Yukawa and his school in that it aimed at resolving dynamical problems remaining in the framework of quantum field theories. As reported by Sakata (1954, p. 35), Tomonaga qualified his attitude as "non-reactionary conservatism" as opposed to Yukawa's revolutionary spirit. This might help render the diversity of approaches.

significant hint, as to how to solve the field reaction problem in general, wherefrom Tomonaga's persistence. As we might recall, since the beginning Tomonaga's reformulation of the nucleon self-field was motivated by the nuclear scattering problem but was meant as a picture that would provide a significant hint to a quantum field theory of elementary particles at large, which could then invariably be applied to electromagnetic or nuclear phenomena.

This epistemological stance, based on the conviction that the problems of the meson theory (and, in fact, of quantum field theory at large) will ultimately be solved within his non-perturbative framework through patient and successive refinements of the approximation method—a constant in the whole of Tomonaga's production—is not endorsed by all of the "Meson Club". Having meanwhile lost his faith in the existence of the meson and loomed over by a "dissatisfaction" with regard to the present state of quantum field theory, Yukawa had gone back to reflect on the problem of the divergent field reaction, which had haunted him ever since his graduation (Yukawa 1991), and began to advocate for a revolutionary change of paradigm. This radical rethinking is in fact not only motivated by the divergences, but also by the lack of manifest relativistic invariance of the current framework of quantum field theory. This is the subject of the talk held at third meson gathering in April '42, entitled "Elementary particles, space-time and causality" (Kawabe 1991).[11] In a notably elegant argumentative style he introduces his proposed solution by drawing a circle—*maru*, with a tacit reference to the act of the calligrapher tracing the archetypical figure of Zen philosophy[12]—on the blackboard, and writing beside it "inseparability of cause and effect" (Takabayasi 1991). Representing a closed hyper-surface in the four-dimensional Minkowsky space, the *maru* represents a finite region where "relativistic causality is no longer valid" and which is vaguely related to the finite space-time extension of an elementary particle. With this device, Yukawa argues, the future theory could more easily be made relativistically invariant and non-divergent. This comes at the cost, however, of giving up the notion of causality, whereby the consequence that cause and effect would no longer be strictly separable (Takabayasi 1991).

---

[11] Yukawa further elaborates on the topic in three papers ("On the foundation of the theory of fields", Kagaku, 12 (1942) 251, 282 and 322 [in Japanese]) which are referred to in a 1943 article by Tomonaga—later republished in English in Tomonaga (1946).

[12] Behind this choice, Yukawa seems to echo an attitude towards the relationship between the Japanese tradition and the West that is evident in two of the most influential intellectuals of the late 19th Century Japan: the philosopher of the aforementioned Kyoto School Nishida Kitaro (1870–1945) and the writer Jun-Ichiro Tanizaki (1886–1965). The former, which figures among Yukawa's influences, rediscusses the Zen tradition in the light of Cartesian and Kantian philosophy; the latter critically discusses the impact of Western technical inventions, and lifestyle, on Japanese culture. In this regard, what Takehiko Takabayasi says about Yukawa's attitude is illustrative (Takabayasi 1991): "[...] Yukawa came to consider, in general, that some type of Oriental thinking might be more suggestive than the ancient Greek and Occidental ideas which up till that time were underlying the development of science".

The Work on the Problem of Relativistic Invariance

While Yukawa's proposal appears in the eyes of Tomonaga as unnecessarily "drastic", the former's insistence on the lack of relativistic invariance of the existing framework of quantum electrodynamics does nonetheless produce the effect of directing his attention to the issue (Tomonaga 1946, originally published in Japanese in 1943). In particular, Tomonaga is called to "the unsatisfactory, unpleasant aspect of the Heisenberg-Pauli theory of having a common time at different space points" (Tomonaga 1966), that is, the fact that space and time are not treated on equal footing in that framework. Despite the contrast between the methodologies of approaching the open problems, and, rather, in virtue of their dialectical opposition, Yukawa's plea thus acts as a stimulator on Tomonaga to instead provide a relativistically meaningful probability amplitude "without being forced to give up the causal way of thinking" or, alternatively said, operating within the *existing* field-theoretical framework. To do this, the strategy is to overcome the limitations of the two dominant existing formulations, Heisenberg-Pauli and Dirac's (Tomonaga 1966). This essentially consists in removing the asymmetry between space and time in the formalism of wave-fields used by Heisenberg and Pauli; and in extending to an infinite number of degrees of freedom (all the space-time points) the particle-based formalism proposed by Dirac ten years earlier in his "many-time theory". Tomonaga's "super-many-time" formalism is outlined already in 1943 (Tomonaga 1946), but its concrete application to electrodynamics and meson theory will take until about 1946 (Hayakawa 1991).

In fact, the obstacles Tomonaga encounters go well beyond the intrinsic intellectual difficulties to become of material order. Starting from the year 1943, the situation of Japan precipitates: air-raids on the city of Tokyo disseminate destruction and fire, and mass drafts among the students and the academic staff take place. Most of Nishina's group is forced to temporarily suspend fundamental research in favour of engineering work on pressing war-related technological concerns. Tomonaga engages, in particular, with the design of waveguides in radar magnetrons and in that context applies theoretical tools from particle physics—like Heisenberg's S-matrix formalism—to resolve engineering problems (Miyajima 1976b; Tomonaga and Miyajima 1976b).

Being located at a Navy base not too far from Tokyo, Tomonaga delivers in the meantime two courses at the Tokyo University of Education—one on quantum mechanics, and the other on nuclear physics and cosmic rays—every other Sunday (the only official vacation days during wartime), which a small group of hungry, metaphorically and literally, students attend. From this pool, he initially selects four students (Ziro Koba, Satio Hayakawa, Hiroshi Fukuda, Yoneji Miyamoto) that would help him develop the super-many-time theory according to a meticulous plan, step-by-step from the simplest to more complex applications. The idea is first to apply the new formalism to the interaction of electrons with the electromagnetic field; then to the electromagnetic interaction of mesons, and on to the strong meson-nucleon interaction (Hayakawa 1991). To prepare his students for the task, Tomonaga instructs them to study selected chapters from Heitler's textbook *Quantum theory of radiation*, the article on cosmic rays by Heisenberg and Hans Euler; and to learn the various relativistic representations of the Coulomb force from the works of Dirac, Fock and Podolsky, Bethe and Fermi, Breit and Möller. At a seminar scheduled on a regular basis, altogether they would first discuss this material, and, then in a latter phase, the advances and problems encountered. Of the several planned, however, only four seminars on the studied material take place over the whole period '44-'45, each happening whenever and wherever the bombing, the hunger, as well as Tomonaga's precarious health, would permit them. Due to these supervening circumstances and the protracting of Tomonaga's low physical and moral condition, the systematic activity on the super-many-time theory begins only some time after the war, in April '46. With a few additional fellows and regaining its pace after the burning defeat of Japan (Kanamori 2016, pp.159–170), the team is able to complete the work in time for the first Meeting of the Physical Society of Japan after the war, in November of the same year.[13]

Reporting on the progress achieved by his team as the last speaker at the Physical Society meeting,[14] Tomonaga underlines that the policy adopted hitherto was "to formulate the existing theory as transparently as possible", but "*without* touching the difficulties of the theory" (Nogami et al. 1947; Hayakawa 1991). To a great extent, the task had been accomplished by exploiting the prescription that Tomonaga had laid out already in 1943 (Tomonaga 1946). This consisted in replacing the non-relativistic partition between a kinematics at the same instant of time and a dynamics fully determining the time evolution—employed by Heisenberg and Pauli and "too much analogous to the ordinary non-relativistic mechanics"—with the partition between "the laws of behaviour of the fields when they are left alone, and [...] those determining the deviation from this behaviour due to interactions" (Tomonaga 1946). This operation, the core of what will be the 'interaction picture', translates into a splitting of the total dynamical equation into a "free" part, describing a particle left "alone" in the vacuum subjected to no external fields, and an "interaction" part which acts modifying that course.

Targeting the Problem of Divergences

While functional to a neat relativistic invariant reformulation of the quantum electrodynamics, the partition devised by Tomonaga did not change the status of the divergences due to the self-energy, which remained included in the interaction part.

---

[13] A chronicle of this intense and somewhat heroic period is kept by Satio Hayakawa. The chronicle reports in detail the steps leading to the formulation of quantum electrodynamics within the super-many-time framework (Hayakawa 1991). Of this first part of the work, for the scope of the present account, we wish to underline Tomonaga's working style. This group of students does not miss the chance to remark on Tomonaga's ability as a director, encouraging leader and teacher, patiently attending to them, and not sparing detailed explanations.

[14] As a further example of the European influence on the basic organisation of research in Japan, these meetings are modelled by initiative of Yoshio Fujioka—upon his return to Japan—after the *Leipziger Vorträge*, informal gatherings organised annually at the University of Leipzig to discuss on some selected topic (Tomonaga and Miyajima 1976b, p. 462).

The old problem of the infinite self-field, common to electromagnetic and nuclear interactions, was indeed left untouched, and therefore would certainly require the "more profound modification" of the new framework that Tomonaga had already anticipated (Tomonaga 1946).

> It is expected that the modification of the theory would possibly be introduced by some revision of the concept of interaction, because we meet no such difficulty when we deal with the non-interacting fields. This revision would then result in the separability of the theory into two sections, one for free fields and the other for interactions [...]. This seems to be implied by the very fact that, when we formulate the quantum field theory in a relativistically satisfactory manner, this way of separation has revealed itself as the fundamental element of the theory.

Assuming this separability, indeed the problem of the self-energy will be solved precisely by deciding where to draw the dividing line between the "free" and "interacting" parts (and accordingly shifting the meaning of the two notions).

With these ideas in mind, the forces of Tomonaga and his group are devoted to the question of self-energy. With the start of the new year, a new cycle of weekly seminars, now held regularly but in a rundown building in the outskirts of Tokyo, begins with the plan to go back to critically review all the literature that had dealt with the scattering of an electron by a fixed external potential and included the influence of self-energy of the electron (Tomonaga 1966; Hayakawa 1991). Those papers— by Braunbeck and Weinmann, Pauli and Fierz, and Dancoff—had roughly adopted either of two strategies: applying a change of variable (a canonical transformation) to obtain the free photons and dressed electrons and to study the scattering between the two non-relativistically; or calculating via perturbation theory the correction to the cross section due to the electron self-energy in a relativistic manner (Dancoff). Both cases had revealed the existence of two types of divergent expressions: one modifying the mass, and a second one related to the scattering process and seemingly irreducible to the former.[15] This implied that a modification of the mass would not be sufficient to eliminate the divergences.

An important hint to the solution emerges as the team repeats the old calculations of Dancoff and realises, in June '47, a mistake in the latter. The rectified expression for the cross section now shows them that the divergence related to the cross section is in fact of the same type as the mass-like one. In other terms, it turns out that the correct expression for the cross section could in fact be decomposed into a finite part plus a divergent term homogeneous to the mass. Ziro Koba and Tomonaga show this crucial result to hold through an intuitive but compelling procedure which they name "self-consistent subtraction" (or "amalgamation method") (Koba and Tomonaga 1947). Adding to the mass of the free electron the contribution of the electromagnetic mass, and subtracting that same term from the interaction energy (so that the total energy is left unmodified); then performing the calculation with such modified interaction, in lieu of the original one, indeed yields the sought-for finite cross section. Within the partition proposed earlier by Tomonaga between the "free" and "interaction" part of the theory, this procedure now sets the former as the free particle *plus* the part of the

---

[15] The expression for the scattering cross section features the integral $\int \frac{dk}{k}$ which diverges logarithmically for high momenta k. For this reason the scattering divergence is dubbed 'ultraviolet'.

interaction responsible for the electromagnetic mass, and the latter as the original interaction *devoid of* the electromagnetic mass. As he would say the year later (Tati and Tomonaga 1948), in this procedure Tomonaga is intuitively guided by an analogy to the method used by Hartree and Fock:

> The procedure in which one includes some part of the effect which would be caused by the perturbation beforehand in the Hamiltonian for the unperturbed system is not new. A well-known example of such a procedure is the Hartree method of the self-consistent field in which one considers the interaction between electrons [in an atomic or molecular shell] as perturbation but some part of its effect is taken into account already in the zero-th approximation, and in such a manner that the perturbation energy has no diagonal elements and, consequently, no level-shift is caused in the first order approximation. The inclusion of the mass-modifying term in the free field equation in our procedure corresponds just to this procedure. We have here considered that the mass-modifying term is already included in the free field equation and, consequently, no level-shift is caused by the interaction between electron and radiation. Our subtraction method is thus "self-consistent" in this sense.

In other words, the self-field, a product of the interactions that any real particle has with the surrounding virtual particles, is seen as analogous to the effect of the real interactions that an electron has on average with the surrounding electrons in an atom, molecule, or solid; and is subtracted from the interaction term according to the same principle. As done for the self-energy of the nucleon (see Sect. 2.1.2), the analogy between real and virtual particles, a solid medium and the vacuum, and more generally between methods in many-body and particle physics, is thus once again invoked by Tomonaga.

The consistent numerical result yielded by this procedure is welcomed with great excitement by the group. Meanwhile, the two American physicists Willis Lamb and Robert Retherford perform a very fine measurement of the energy spectrum of the hydrogen atom by means of a new technique and discover a splitting between two orbital levels that, according to the first-order corrections, should be degenerate in energy (Lamb and Retherford 1947). As Hans Bethe shortly afterwards shows plausibly through a back-of-the-envelope calculation, the observed energy difference is a consequence of the field reaction being slightly different for the two orbitals, whose measurable effect is now directly experimentally detected for the first time (Schweber 2020). The news reaches Japan only some time later, almost by chance, through a column in the popular magazine *Newsweek* (Hayakawa 1991).[16] Tomonaga discusses the finding with the group, and with the help of Takao Tati and Ziro Koba, they first derive more rigorously Bethe's results by applying the canonical transformation of Pauli and Fierz; and then provide a relativistic generalization of it within the framework of his super-many-time theory (Tati and Tomonaga 1948; Tomonaga and Oppenheimer 1948). Through this same generalised transformation,

---

[16] During war time, no international literature, whether popular or scientific, can be found in Japan. After the war, with the establishment of the so-called Centres for International Exchange (CIE) by the Americans, international journals and magazines were made available with some delay (see, e.g., Kinoshita 1991). The fact that the news of the Lamb shift was not apprehended through the published scientific sources, though, testifies to a state of relative isolation of the scientific community persisting in these first few post-war years.

they are able to give formal ground to the intuitive self-consistent subtraction that had been presented in a preliminary form in 1947 (Iwata et al. 1948). So formalised, the procedure will allow for the shift of all the electron-related divergences from the interaction part into the free part of the theory, with a "mere" modification of the electron's mass and charge. In such a so-called 'renormalised' theory, the quantities that used to yield nonsensical divergent predictions—the level shift and the radiative corrections to the scattering processes—turn out finite, and are thus in principle experimentally adequate.

Before being written down in a number of articles, these results are presented in a preliminary way by several speakers at the Annual Meeting of the Japanese Physical Society, in fall '47, where there is a pervasive presence of Tomonaga's group, with roughly a third of all the talks (Iwata et al. 1948). There, the work on renormalisation stands out as prominent, and testifies to the technical and organisational triumph of his approach over Yukawa's. Nambu, who presents a phenomenological explanation of the Lamb shift, would later comment on how "awed" he felt before Tomonaga's resolution (Nambu 1991).

With the framework of renormalisation conceptually laid out, the work of the ensuing year sees a continued team work to actually refine the calculations of the Lamb shift and the radiative corrections to various other scattering processes. Nambu, who had come back from the war when the group had already formed, hitherto a spectator of the group's activity through Koba's work, is put to independently perform the calculation of the Lamb shift with an alternative method to that used by Fukuda, Miyamoto, and Tomonaga (Nambu 1949). The work of Nambu refines the numerical result given by the three physicists. But it also provides a critical discussion of the approximations adopted and a lucid comparison between his and their method, betraying a full familiarity with the renormalisation procedure. As we will see, this expertise, alongside the analogical correspondence between real and virtual particles and its combination with the Hartree-Fock method employed in the context of renormalisation, will have a crucial role in the discovery of spontaneous symmetry breaking.

2.1.4   The Elementary Particle in the Aftermath of Renormalisation

The quantum electrodynamics developed by Tomonaga—and independently by Schwinger and Feynman, as is well-known—is indisputably a computational and experimental success: at once, it casts electrodynamics into a quantum field theory, where both electrons and photons are treated as relativistic quantum fields and made to mutually interact. It also resolves the problems of the infinities that had threatened quantum electrodynamics since its early formulations in the late '20s by organising and "explaining" them. The theory so renormalised is able to predict with great numerical accuracy the finest details of the hydrogen spectra, as well as the scattering phenomena. This positive outcome definitively reaffirms, in the years 1948–49, a paradigm in which the concept of the particle, properly extended, retains its essential validity in the new context of the quantum field theory (Blum and Joas 2016).

This extended concept of particle has an implication that is relevant for our story and which we wish to make here a brief digression on. In the new scheme, the electron that we observe, with its experimentally measurable mass and charge, is the *dressed* electron: it is the result of the *bare* electron—with a different, non-zero (in fact, infinite) initial mass and charge—plus the "cloud" of virtual photons making up its self-energy, and can never be observed in its bare form. In more specific terms, the bare electron, thought of as not interacting with the photon field, is now the first elemental (the zero-th order) but "noumenal" object in a perturbative series. In that series, each successive component represents an additional, smaller but increasingly complex, interaction with the surrounding virtual photon cloud *modifying* the bare mass and charge. The sum of these terms cumulatively gives rise to the "phenomenal" object (the dressed particle) with the observed values of mass and charge. If the observed particle can be seen as the result of the bare particle interacting with the vacuum, in analogy to a particle in a dielectric solid, in practice there is an essential disanalogy which Tomonaga points out. While for the particle in the dielectric there exists an experimental procedure to probe both its bare properties (outside of the dielectric) and the "renormalised" properties (inside the dielectric), this is not possible for the particle in vacuum.[17] As we elaborate on in Chap. 5 and Sect. 5.1.2, the epistemic gain of the analogy that Nambu will draw lies in this disanalogy.

In the ecosystem of theories of elementary particles that emerges in the late '40s, this concept of particle promoted by renormalised quantum electrodynamics provokes mixed reactions. On the one hand, the impetus of the success tends to set electrodynamics as the paradigmatic model for the other field theories. This leads various theoretical physicists to a re-examination of the difficulties in meson theory in the light of the new model, and in general, as was the original intent of Tomonaga, to extend mass and charge renormalisation to the meson-nucleon interactions (see, e.g., Kanesawa and Tomonaga 1948; Nambu 1948; Van Hove 1949; Watson and Lepore 1949). All in all satisfied with the ontology of renormalisation, these programs devote themselves to understanding whether the scheme of quantum electrodynamics is valid beyond electrons and electromagnetic fields, and what of this paradigm is extendible to other fields and particles (Rohrlich 1950; Matthews and Salam 1951). A different question characterises those who instead recognise in renormalisation the status of a transitory and provisional patch—however effective—which thus awaits a radical solution to the divergence problem. As was the case for Yukawa, the question there is in general whether the divergence difficulties are related in some way yet

---

[17] In Tomonaga's words (Tomonaga 1976b, originally published in 1949): "In a dielectric substance it is possible to distinguish between E and D by boring a small hole in the substance and measuring the force acting on a test-body there, but it is impossible to bore a hole in a vacuum. Although it must be due to a defect of the theory that such an answer as infinity should be obtained, there is no experimental procedure, in the case of these simple phenomena [renormalisation of mass and charge] to determine what finite value ought, actually, to appear.

to be determined to the structure of elementary particles, and therefore should be solvable through a scrutiny of the more fundamental assumptions underlying field theories.

Originally established when cosmic rays were the only source of high-energy particles, both philosophies are influenced and affected in their course by the advent of the increasingly powerful particle accelerators going on-line in this post-war period, with the accelerator in Berkeley (California) ahead of all. As the first artificial mesons become available in such facilities, theorists are pushed in an effort to derive the observable consequences of the proposed models in order to test and further the theoretical achievements. Soon to emerge from the accelerators "plucking" the quantum vacuum more and more forcefully are not only the known particles, but a growing number of other species. This lead back, from another direction, to the question of the structure of an elementary particle, their elementary and composite character (see Section "Are elementary particles really elementary?" below).

In this complex of circumstances evolving in the aftermath of the new electrodynamics, Tomonaga's position is one of fundamental seeking, as it appears from the research interests and the correspondence of those years (Tomonaga 1976c). He is receptive to the further developments of the renormalisation programs and hopeful for the results of the accelerators. Yet, as we will see in the next section, he grows disenchanted by renormalisation, and expresses the need for a rethinking of the basic assumptions underlying the framework of field theory, and the concept of the particle lying at its base.

Rethinking the Concept of Elementary Particle: Tomonaga's Testament

In a nodal paper from January 1949, issued in the magazine *Kagaku*,[18] and eloquently entitled "The development of elementary particle theory-Discussions centring on the divergence difficulties", Tomonaga critically looks back at his own contribution and what had been accomplished so far, and forward to where to hypothetically go (Tomonaga 1976b). The discussion has the tone of a confession, a testament, which sets the ground by comparing the fundamental problem of divergences in quantum field theory—which, according to him, his renormalisation procedure had only patched[19]—to the problem of divergences of the cavity radiation, a problem which "supplied the motive for the discovery of quantum theory". After a precise, concise, but accessible critical discussion of the present treatment of divergences in quantum electrodynamics, he reflects on the characteristics of the theory:

---

[18] *Kagaku*, commonly translated as *Science*, is a beautifully illustrated monthly magazine with a very broad scope. It features quite technical articles ranging from mathematical logic to disaster prevention, to physics, biochemistry, and various Japanese translations of international papers.

[19] Quite significantly, he writes: "When a theory [quantum field theory in its present status] is incompetent in part, it is a common procedure to rely on experiments for that part [...] our method has brought the possibility that the theory will lead to finite results by the renormalisation even if it contains defects".

> [...] when, in performing a theoretical calculation for certain phenomena [...], we assume the interaction to be [non-existent or] small and carry our calculations out no further than the first approximation in which this phenomena actually appears, then the results given by our theory is such that it agrees sufficiently well with experiments. [...] But when, in view of the fact that they are not perfect in that they do not take the effects of the interaction between the electron field and the photon field into account, we try out calculations [...] regarding these effects as corrections, these corrections always turn out to be infinitely large, so that the agreements of the results with the experiments, far from being improved, are utterly destroyed.

To this state of affairs, Tomonaga maintains, he and his group have reacted by "chasing these infinity-difficulties [the mass and the charge of the free electron], as it were, step by step, into a corner" and, as a result, "have found that the theory, imperfect though it is, is still usable to a considerable extent". This is not, though, in Tomonaga's eyes, more than a "transitionary expedition" where it is hard to say if the present theory "is of sufficient value to act as something of a guide, or it is perfectly useless". In other words, that was a progress—and here Tomonaga belatedly echoes Yukawa's in the tones—that did not extinguished a fundamental question, and demanded for an equally fundamental conceptual revision:

> What, then, are we going to do about the infinities [...]? No one can say for certain at present. [...] The solution might involve a fundamental change in our concept of nature, such as the discontinuous structure of time and space, or it might require a radical modification of our concept of elementary particles and their mutual interactions.

Pursuing the difficulties down into an internal conflict between basic concepts (Taketani 1971) which is reminiscent of Yukawa and Taketani's methodology, Tomonaga questions the limits of applicability of the concept of particle and the distinction (see Sect. 2.1.3) which had brought him to the relativistically invariant field theory in the first place:

> It would be only when no interaction exists at all, or at most one which can be made arbitrarily small between elementary particles, that we arrive at the concept of individual elementary particles. But the conclusion from our theory, far from justifying this, has it that the effects of the interaction are always infinitely large. Consequently, our theory destroys, in its conclusion, its basic assumption. It thus involves an inconsistency.

In what seems, *mutatis mutandis*, to be a re-enactment of the fundamental dilemma that plagued the quantum theory of fields in the early '30s (Blum and Joas 2016), physicists are now before a theory that, now able to perform calculations not directly frustrated by the infinities, provides extremely accurate corrections to the old quantum theory. Yet, its efficacy is "unreasonable" due to, or at least blemished by, a basic inconsistency that points beyond the concept of individual particle. This self-inconsistency, as Tomonaga comments, "evidently" arises as a consequence of the abuse of the classical-to-quantum analogy:

> [...] we have borrowed the two concepts of field and the interaction between them, *per se*, from the classical (unquantised) field theory. In an unquantised theory, we can actually make the interaction between various fields arbitrarily small. This is because an infinitely weak field is conceivable in classical theory. From these points, concepts such as a field independent of others, and the interactions between them were, in classical theory, proper

concepts with which to describe nature. But the result of the quantization is, as said above, inevitable to cause the interaction to become infinite. From this viewpoint, it seems that, in the current theory, such concepts as a field independent of others on the one hand, and the interaction on the other require fundamental modification [...].

Assumed a priori, the classically-derived concept of individual particle is a posteriori not justified by the weakness of the effect of the mutual interactions, which de facto invalidate it. The contention is whether in a quantum theory a truly non-interacting particle, even when properly modified through renormalisation, is in fact a too idealised, if not invalid, starting point, both conceptually and technically. Since, as we have seen, the individual particle is the ontological base of the perturbation method, Tomonaga's criticism automatically calls into question the latter.

The problematic presence of infinities in the mass and charge of the electron would alone not be able to justify the rather radical position we have just examined. Tomonaga's position can instead be understood by considering that he and much of the Japanese community working on particle physics, for the historical reasons we have given previously,[20] seek for a theoretical framework that could include also, and especially, nuclear phenomena. In light of this aim, the criticism of the concept of individual particle is linked to the structural inadequacy of perturbation methods, which can treat only weak forces and thus "few-body" problems. The same criticism advanced by Tomonaga in 1939–40, which animated his Hartree-Fock non-perturbative model of the nucleon, reoccurs now. To a similar judgement will correspond a similar reaction: Tomonaga seeks a possible way out by turning to many-body physics, structured instead to deal with processes non-perturbative in nature, in which multiple collisions and stronger interactions need to be accounted for. As we will see in Sect. 2.2.1, this shift will mature during the one-year sojourn—for the two terms 1949–50—at the Institute of Advanced Studies in Princeton, USA, where Tomonaga will end up investigating the collective excitations of an electron gas, the "hydrogen atom" of solid-state many-body systems.

Are Elementary Particles Really Elementary?

With Princeton as a final destination, in the summer of 1949 Tomonaga sets sail for San Francisco from Yokohama. Through the correspondence addressed to his collaborators in Japan,[21] we can follow the encounters, interests, impressions and struggles of his pilgrimage to the temples of theoretical physics in America. "I went to Berkeley today and met Serber, Weisskopf, Wick, etc., and had them show me

---

[20] Reasons of pride are also an important factor for the constant interest of Japanese physicists in the nuclear phenomena. This can be understood as a result of the status and fame that Yukawa had in Japan, even before the Nobel Prize (see, e.g., Kanamori 2016).

[21] These letters were published in the informal particle physics journal *Soryushiron Kenkyu* since their technical content was thought to be of high interest to the Japanese community of theorists, whose only link to the forefront of research at the time is often the correspondence from outposts abroad. A selection of them can be found in Tomonaga and Miyajima (1976b).

the cyclotron" breaks off the first letter from San Francisco (29th August 1949), and goes on to technically describe the famous accelerator (then the most powerful in the world and completed in 1947), interspersing admiration with jests:

> It is like going to practice with a chick first and then make it real. Unfortunately, I couldn't understand Serber's explanation because I didn't understand a word of English. Serber is a very kind person with a face similar to Mr. Sagane [a Japanese fellow working at the accelerator in Berkeley]. If this is completed in two years, it may be possible to create a negative proton. The energy is 6 GeV or 2 GeV (Kodaira heard 6 and Tomonaga heard 2!)

On this, follows the reporting of the University of Chicago and the encounter with Enrico Fermi [in italic the verbatim words from Fermi's paper]:

> I wrote the story of Berkeley in a letter to Fujimoto, Miyamoto and Fukuda earlier, so let's talk about Chicago now. I met Fermi in Chicago. It's not a big Italian guy, it's an uncle whose face is dark, and he talks about strange things and laughs happily. One of his current activities is asking "Are Mesons Elementary Particles?" (I sent the Manuscript to Miyazima) [...]. *In recent years several new particles have been discovered which are currently assumed to be "elementary" that is, essentially, structureless. The probability that all such particles should be really elementary becomes less and less as their number increases.* It's a good thing. [...] He says that theoretical physicists are all too much thinking the same way. I thought this saying was Fermi spot on. What are your thoughts?

What Tomonaga here quotes from is a manuscript just submitted to the Physical Review (Fermi and Yang 1949) written by Fermi and Chen-Ning Yang—a young promising Chinese theoretician about to leave for Princeton—entertaining the possibility that mesons, rather than being elementary, are composed of a nucleon and an anti-nucleon. Beyond the specific application, admittedly "crude", the authors wish to provide one "illustration of a possible program of the theory of particles" whose general philosophy is that some of the known particles, while manifesting as if elementary, might in reality be bound combinations of elementary particles—forming due to a yet unknown attractive interaction—much like molecules are combinations of a finite set of atomic species.

The model of Fermi and Yang is but one of the examples of a soul-searching enquiry on the fundamental structure of matter that characterises this period of particle physics, emerging as a reaction to the proliferation of particle species out of the increasingly powerful accelerators, from the dissatisfaction with regard to the divergence problems, or from a mixture of both. While the model by Fermi and Yang can be ascribed neatly to the first kind, and, for instance, the theory of non-local fields proposed by Yukawa at the same time (Yukawa 1950) definitely belongs to the second kind,[22] parallel theoretical attempts arise which aim at constructing the observed particles and their properties as non-perturbative bound states of a *single* or a few underlying "ur-fields" (unobservable in isolation).[23] In doing so, these proposals seek to settle the two issues—variety of particles and divergences—at once

---

[22] This theory is the result of the speculations of Yukawa that we have mentioned in Sect. 2.1.3 (see, e.g., Takabayasi 1991).

[23] Here we refer to the generically called non-linear models proposed in those years by, e.g., Fritz Bopp (1948) and Heisenberg (1950). For a detailed account of the latter, see Blum (2019).

by taking the monistic principle (*reductio ad unum*) that is latent in the Fermi-Yang program to its extreme consequences. As we will see, much of the direction taken by Tomonaga, and the model of elementary particles obtained a decade later by Nambu can be understood as a synthesis of the markedly realistic character of the Fermi-Yang composite model with the aim of the more speculative proposals.

The conjuncture at Princeton and the work that Tomonaga conducted there on collective phenomena are an important veer, and constitute a central element in the development of spontaneous symmetry breaking. This is the subject of the next section.

2.2 The Collective Description of Many-body Systems

It is an indisputable fact that the course of physics in the United States has been positively affected, from the 1930s on, by the massive emigration of scientists from fascist dictatorships in Europe. In fact, the phenomenon is not restricted only to the pre-war period nor to Europeans, but rather continues for quite some years after the war and features for example, as we will see, an influx of Japanese physicists escaping the hard conditions of the post-war Japan. The Institute for Advanced Studies at Princeton is able to profit exceptionally from this conjuncture strategically intercepting from the early '30s the triad of Albert Einstein, Hermann Weyl, Kurt Gödel, and other personalities in the course of the subsequent years which have increasingly endowed the school with an immense prestige (Leitch 2015). In the midst of the general reconversion of war-related research back to fundamental issues that had steered physicists in the pre-war days (Hoddeson et al. 1992), the Institute's trustees make another far-sighted move: they offer the directorship to Robert Oppenheimer, who, back from the research leadership of the Manhattan project and no longer inclined to teaching, takes Princeton to its full power as a research coordinator (Bird and Sherwin 2007). Browsing through reports, seminars, published papers, as well as the list of visiting scholars in both the schools of mathematics, and historical studies,[24] it is hard to imagine a denser place for a theoretical physicist to reside for a temporary stay. A report of the burgeoning term '48-'53 carefully drafted by Oppenheimer himself provides an eloquent overview of the vocation, with bursts of restrained excitement (Oppenheimer 1954):

> In the years 1948–1953, there has been a marked growth in the work of the Institute in theoretical physics. This was a natural development for the Director to undertake. [...] The Institute has always had a few members interested in theoretical physics; this was inevitable in a school of which Einstein was a first member, and which Pauli and Bohr visited from time to time. But both the scope and the character of the work have changed. We have appointed two professors, Freeman Dyson and Abraham Pais; we have made arrangements for a continuing or recurrent association with the Institute for a number of physicists: Niels

---

[24] A number of research reports, letters, seminars, list of visiting scholars and the articles resulting from the work carried out at the Institute are publicly available on the web page of the Princeton Archives.

Bohr, P. A. M. Dirac, Res Jost, G. Placzek, W. Pauli, L. Van Hove and C. N. Yang. The number of annual members has increased rapidly; in 1953, we had about 25 men in this field working at the Institute; and in the whole period, we have had about a hundred such members. Their work has eventuated in the publication of between one and two hundred papers in the Physical Review [...].

Speaking in further detail about the invited temporary members, Oppenheimer underlines the strategy behind the composition and the diplomatic action carried out leveraging the large international network constructed in the past[25]:

The members in physics come to us from many places, in many different stages of their work. A few of the most brilliant post-doctoral students in the country come here for a year or two of further study; a few come to us on leave from assignments involving serious administrative duties with the government or with industrial laboratories, and have an opportunity both to complete work which they have wished to undertake, and to learn of the new things that are under development. Many are scientists from abroad. In fact, it was in physics that we first re-established effective and continuing contact with the Japanese scientists. We have had a few brilliant young Chinese, who have immigrated from China after the war. Members come from India and Pakistan, and from almost every country of Western Europe.

Among these prominent temporary members, justifying the emphasis on the reconnection to the Japanese scientific community at large,[26] figure Yukawa and Tomonaga, respectively visiting the Institute in '48 and '49. Second in the long list of eager applicants—which includes Nambu too—Tomonaga is invited to join, in Oppenheimer's words to General MacArthur (February 20, 1949), "to enable him to resume cooperative studies with American workers in the field [of fundamental theoretical physics] [...] profitable not only for Dr. Tomonaga and his colleagues but for our group here" (Oppenheimer 1949). And it is to him that the "leave from assignments involving serious administrative duties" is, in particular, applicable. Besieged by educational and administrative chores back in Japan, Tomonaga regards the sojourn at the Institute as a respite from the lecturing and the running of his department,[27] and arrives in Princeton indeed with the aim of learning about the new developments in the field of particle physics.

---

[25] In a letter to his mother, Freeman Dyson, one of the makers of quantum electrodynamics, writes (Dyson 2018): "As a physicist, Oppy [Oppenheimer] has his limitations, but as a politician he is really outstanding".

[26] For various reasons—two being Oppenheimer's fascination with the Far East and the wish to make reparations for the atomic bombs—the relation between Oppenheimer and the Japanese physicists is significant. Oppenheimer, in a short note written with Robert Serber on the Physical Review (issued in 1937) had signalled for the first time the virtually unknown work of Yukawa to the Western physicist (Oppenheimer and Serber 1937), and would later invite—acting as an interceder—Tomonaga to report on the same journal the independent and practically unknown progress done in Japan by Tomonaga and his group in the domain of quantum electrodynamics (Tomonaga and Oppenheimer 1948). This communication is the first appearance of the work of the Japanese group in a Western journal.

[27] This is indirectly confirmed by a letter of Dyson to his mother: "During this year he says he has not done a lot of work, he has used his time at Princeton as a holiday, and has greatly enjoyed the freedom from lecturing and running his department. Of course he is in Japan besieged with research students clamouring for help and advice, even more than I am in England." (Dyson 2018).

It is indeed for him a period to meet for the first time and connect to physicists working on a variety of topics, whom he had only pictured through the paper. Beside the impressionistic portraits of the personal encounters and conversations with Richard Feynman, Freeman Dyson, Wolfgang Pauli, and other more or less renowned theoreticians of the time, the letters to his fellows in Japan return technically detailed sketches of the advances, problems encountered and open questions in various undergoing theoretical programs (Tomonaga 1976c). Judging from the themes dwelt on in the correspondence, Tomonaga's focus appears to be on those aforementioned attempts at deriving particles out of underlying fields interacting non-linearly, or as compounds of other particles along the line proposed by Fermi and Yang. In light of the position expressed a few months earlier in *Kagaku*, on the one hand, this confirms the shift of interest away from renormalisation towards a more radical conceptual revision. In the case of Tomonaga, such a revision remains rather pragmatic and anchored to a mathematical method: it consists in a rediscussion of the picture of bound nucleons in nuclei from a more fundamental perspective. The underlying idea is to find a framework to replace the (hitherto inadequate) description of strongly interacting fermions with one in terms of bosons (Tomonaga 1950). Attained by drawing, once again, from the reservoir of many-body physics (as he had done with the Hartree-Fock method), this work involves a shift from the individual particle description to the so-called collective description. A similar shift is independently and coincidentally advocated by David Bohm as a result of his work on the gas of interacting electrons and ions, called plasma. With their own languages, approaches and conclusions, both works will have a significant conceptual impact on many-body physics at large and, through that, on the discovery of spontaneous symmetry breaking in particular. In the next few sections we dwell on the paths that conducted Tomonaga and Bohm to their collective descriptions.

2.2.1 A Change of Paradigm in the Description of Nuclei and Plasmas

From the late '20s and the advent of the new quantum mechanics, the theoretical and experimental activity in many-body physics had been going on in parallel with the search for suitable quantum field theories that we have hitherto followed. In a certain sense, less "fundamental" than the latter, the physics of many bodies consisted for a large part in the application of the notions of quantum mechanics to many-body systems, from the complex nuclei to the multi-electron atoms, molecules, and solids. Partly owing to the growing technological relevance of its applications, ranging from the vacuum tubes to radars to the thermonuclear reactions, this branch of physics had been steadily growing in the pre-war period and quite significantly during the war (Hoddeson et al. 1992; Weart 1992). This technological interest, the increased investments, specialisation and manpower prepare the ground for a golden decade of many-body physics after WWII—and of solid-state physics in particular—where a few key conceptual advances open breaches to the rationalisation of several phenomena. One such breach is opened in the stronghold of the single-particle models by the concepts of collective phenomenon and then, as we will see, spontaneous symmetry breaking. All variations on the same theme, the single-particle models had been made the preferred tool, and in fact practically the only available one, to treat complex systems as diverse as atoms, molecules, solids and nuclei under the simplifying hypotheses that all the particles move in one common potential where the inter-particle interactions enter only as averages, and correlations are considered at most perturbatively.

Recognised as too rough approximations especially for nuclei and solids already at the end of the '30s—among others, by Bloch (1933), Bardeen (1936b), Bohr (1936), Bethe and Bacher (1937)—these methods, refined in several ways, nonetheless continue to be used sustainedly and (surprisingly) satisfactorily in the pre-war period, as well as applied to the numerous war-related issues which required quick, workable and computationally inexpensive numerical solutions.[28] As post-war research goes back to less pressing problems, however, it becomes more and more obvious that the single-particle approach hits a limit for a class of die-hard many-body phenomena to which the inter-particle interactions and the dynamical correlations between the various bodies are suspected to be essential. And while such is believed to be the case in a superconductor, as we will see in Chap. 4, it also already occurs in heavy nuclei, and the comparatively simpler plasmas or metals, where, due to the high density and/or electrostatic interaction between the many particles, the motion of one particle is transmitted to all the others—each particle is said to be *correlated* with *all* the others—giving rise to phenomena that simply cannot be captured starting from single-particle solutions/levels alone (as both standard perturbative and non-perturbative methods do) and instead demand to take into account its complement, the collective description. Precisely these issues are systematically investigated starting from the late '40s, and targeted with sufficient generality from two different directions by Tomonaga and Bohm.

Tomonaga and the Many-Body Problem in Nuclei: Inspirations from Solids

Among the multiple applications of the single-particle paradigm to nuclear physics in the '30s, the authoritative voice of Niels Bohr struck a discordant note. This voice had quite an influence on Tomonaga's path to the collective model and it is worthwhile to briefly recall it.

As Bohr argued, the recent data from the systematic investigation on the radioactivity induced by fast neutron bombardment of heavy nuclei—conducted by Enrico

---

[28] Prominent examples are the calculations of the energy needed to extract an electron in a solid—the so-called work function—and the stable configuration of a high-density electron gas, respectively employed, for instance, in the construction of the transistor and the modelling of thermonuclear reactions.

Fermi and collaborators over the course of the years 1934–35 (Amaldi 2004)—suggested that the intense energy exchanges between the individual, densely-packed nucleons were to be considered an essential part of the nuclear phenomena. This meant that the single-particle approximation, "so effective in the former [atomic] case, loses any validity in the case of the nucleus where we, from the very beginning have to do with essential collective aspects of the interplay between the constituent particles" (Bohr 1936). To corroborate this view there stood the fact that the excitation spectra of the heavy nuclei in question "closely resembled" those of "solid bodies", "elastic substances", or "a sphere under the influence of a [...] surface tension" (wherefrom the name "liquid drop" model). Leveraging quite deftly the analogy with these physical systems, Bohr convincingly concluded that the excited state observed by Fermi was to be attributed to "some quantised collective type of motion of all the nuclear particles" rather than "to an elevated quantum state of an individual particle [i.e., a state where one particle only is excited to a higher level]", in contrast to the interpretation offered by the single-particle models (Bohr and Kalckar 1937).

As we have seen, Tomonaga had tentatively applied himself to the problem of nuclear potential since the early '30s; and by the time Bohr's conclusions were drawn, he and Kobayashi were working to take into account nucleon-nucleon correlations in a perturbative way starting from the Thomas-Fermi single-particle approximation (Tomonaga 1968).[29] Apprehending from Bohr's work what was seemingly the utter inadequacy of such approximations to treat nuclear systems, Tomonaga embraced the collective approach and set out to give it a more rigorous treatment (Tomonaga 1938). Assuming the nucleus as a gas of fermions—considering thus the nucleons as hard spheres interacting only via exchange interaction—he uses the framework of kinetic theory to compute the hydrodynamic properties of the "nuclear matter", its viscosity and heat conductivity. Tomonaga's "microscopic" model essentially confirmed the theoretical possibility of collective oscillations, though in practice the nuclear matter was found to be so viscous that any oscillation, according to his model, turned rapidly into a "completely irregular heat movement". This conclusion allowed Tomonaga to assert, *per viam negationis*, that responsible for the resonances observed in the experiments cannot "simply be elastic vibrations of the Fermi gas sphere" but must be (our emphasis) "obviously internal vibrations that are linked to the *deviation* of the nuclear matter from the gas state and indicate a liquid-like structure that is not included in our simplified model."[30]

Due to these limitations and its unrealistic character, this first crude collective model would soon after be abandoned by Tomonaga in favour of the work on what would become the intermediate coupling model of the nucleon.[31] It is now, in 1949,

---

[29] As reported in the same recollection, the two Japanese physicists were, in fact, anticipated by Heisenberg's collaborator Hans Euler who had been independently working on the same issue and already published his results. Tomonaga learns of the fact upon his arrival in Leipzig where Euler was working. Beyond the fact itself, this signals once again the isolation of the Japanese scientific community.

[30] Author's translation from German.

[31] This model, as we have detailed in Sect. 2.1.2, pioneered the application of the Hartree method to the meson cloud surrounding the nucleon, and thus marked his return to the single-particle approach.

when the larger theoretical horizon invites a microscopic, field theoretical re-framing of the nuclear problem, that Tomonaga returns to his collective model. While reincarnated in a different mathematical framework as compared to then, the fundamental idea remains essentially the same: the coupling between the individual particles in nuclear matter is so strong that their effective motion cannot be regarded as independent and uncorrelated, nor as *almost* so by accounting for the correlations perturbatively (Tomonaga 1950). As a consequence of this, Tomonaga observes, recalling Bohr's conclusion, the excited states of the atomic nucleus can by no means be considered as excitations of a single particle in the mean field produced by the others. In other words, in such an assembly of particles, assuming the existence of one-particle "levels" and excitations is unjustified both as such and when used as the starting point (zero-th approximation) of a perturbation series—which in turn would be justified only if the inter-particle coupling were small. Differently from the statistical approach used more than a decade earlier, the solution outlined by Tomonaga now has a much larger breadth and generality. It consists in transferring to the nuclear matter, and generalising, an approximation method developed by Felix Bloch in 1933–34 (Bloch 1933, 1934) by analogising the excitations produced by a particle traversing a charged medium—and losing energy via a mechanism called "stopping power"—to the ones produced by a neutron traversing nuclear matter.

"In his famous work on the stopping power of charged particles" Tomonaga remarks in the introduction to the 1950 paper, "Bloch has treated the excited states of the Fermi gas not as states with holes and excited particles, but as states in which the gas [as a whole] oscillates" (Tomonaga 1950). When quantised, Bloch's same approach correspondingly revealed that the density fluctuations of such an assembly of non-interacting fermions can in fact be equivalently described as quantised sound waves. Starting from this result and continuing along the line prefigured in his former kinetic model, Tomonaga constructs a generalisation of it, in one dimension, in which the Fermi gas is made a "liquid"—that is, where free particles are replaced by interacting ones—and proves that Bloch's method is valid even in the presence of an interparticle interaction, however strong.[32] By doing so, it is realised that an assembly of interacting particles not only *can* be described by a non-interacting assembly of collective excitations[33]—wave-like oscillations of the whole system of particles—but *must* be so described: because each excitation, as originally intuited by Bohr, is actually seen to correspond to a very complicated superposition of single-

---

[32] The approximation fails only in case the range of the force is "too short". See Tomonaga (1950) for the specifications.

[33] Put in simple terms, this is achieved by replacing the particles' variables $\psi(x)$ (fermions) by the associated particle *density* variables $\rho(x) = \bar{\psi}(x)\psi(x)$ (bosons), which are instead the variables for the sound field. In this way, while the former representation becomes non-linear in the presence of an inter-particle interaction—which has the bilinear form $\bar{\psi}\psi\bar{\psi}\psi$—the problem remains linear in the latter case (in which it has the form $\bar{\rho}\rho$). With this, the non-linearity that renders problematic the treatment of interacting fermion systems is circumvented; in this sense, the "sound approximation" is a linearisation of the otherwise non-linear equations of the system. This sort of linearisation procedure applied to otherwise non-linear many-body problems will also characterise the approximation methods later adopted by Nambu.

particle states, the latter alone are simply inadequate when dealing with a strongly correlated interacting system of particles, be it an atomic nucleus, a plasma, or a solid. Highly problematic when described from the single-particle perspective (as introducing complicated non-linear effects), when reckoned instead through the collective representation, the inter-particle potential shows up merely as an extra term in the spectrum for the collective oscillations, transforming it from that of a regular wave—with the energy going to zero for infinite wavelength—to one having a gap.[34] In other words, the interaction potential acts as if endowing the collective boson with a "mass" proportional to the magnitude of the interaction. One can therefore say that switching on long-range interactions between the fermions translates into a mass in the equivalent boson description.

As we will see at the end of Chap. 4, years later this mechanism by which (gauge) bosons acquire mass out of an underlying interaction will be recognised at work in superconductors too and, from the latter, it will then be transferred by analogy in the realm of elementary particles.

2.2.2 Bohm and the Many-body Problem in Plasmas and Metals: Inspirations from Quantum Electrodynamics

From a rather different direction and with correspondingly different purposes,[35] results similar to Tomonaga's are obtained by Bohm in collaboration with his two students in Princeton, Eugene Gross and David Pines. Having refused to join the Manhattan Project at Los Alamos (Freire 2019), starting from 1943 Bohm had dedicated his graduate period at Berkeley to studying a state of matter called plasma

---

[34] In the case of a gas of electrons interacting via long-range Coulomb forces (the plasma), this gap corresponds to the so-called "plasma frequency" and to the mass of collective excitations called plasmons.

[35] As surprising as it may seem, the way Bohm came to treat plasmas differed from Tomonaga not only on the technical level, but in that it had a "political" charge. As recollected by Bohm in a famous interview with Lillian Hoddeson in 1981 (Hoddeson 1999): "First of all, it was a sort of autonomous medium; it determined its own conditions, it had its own movements which were self-determined, and it had the effect that you had collective movement, but all the individuals would contribute to the collective and at the same time have their own autonomy." As noted in Freire (2019) "Political metaphors, philosophical views, and physics problems were all part of the scientific reasoning of a creative mind as Bohm's. [...] Bohm's work on plasma coalesced a scientific style to be further developed throughout his life. He was inspired by daring analogies, often drawing together meanings from distant domains of experience." More on this can be found in the quoted source (pp. 41–42).

and its behaviour when subject to magnetic fields.[36] Such studies, directed by the Radiation Laboratory, were part of a war-related nuclear program and aimed at the enrichment of uranium, a process achieved through extracting its lighter isotopes by applying a magnetic field to the hot uranium plasma. In moving to Princeton University in late '46, under his first appointment as assistant professor, Bohm is determined to continue his research on the subject, setting up a more conceptual, wider, and less purpose-driven program than before, right at a time when the war-related interest on plasmas is subsiding (Hoddeson et al. 1992). Regarding plasma as a peculiar state of matter on its own, the program sets out to study various aspects of the individual and collective motion of its particles—mostly done in collaboration with Gross— and then to use plasma as a paradigm of the behaviour of other solid media (Gross 1991)—a work Bohm will conduct with Pines. In particular, Bohm aims at addressing phenomena in metals that were not yet satisfactorily understood on the basis of the single-particle model, and possibly also the still exotic phenomenon of superconductivity. This strategy of "recognising" plasma behaviour in solid matter by analogy, while failing to explain superconductivity—for which a crucial piece was still missing—will have significant implications for solid-state theory and particle physics.

While certainly devoted primarily to low-energy physics in a landscape where a large part of post-war theoretical physics at the forefront is fidgeting with the divergence problems in quantum electrodynamics and meson theory or is in search of a unified theory of elementary particles, Bohm is not indifferent to such problems, but likely resistant to their existing solutions (Hoddeson 1999; Freire 2019; Gross 1991). Not dissimilarly to Tomonaga's in that respect, his choice also betrays the search for a different paradigm, alternative to the existing ones, which could eventually address those problems more radically. In particular, turning to the physics of plasmas could offer, in his view, the chance to arrive at a reconciliation of the collective and individual descriptions—the wave and particle aspects—that had characterised quantum mechanics since its inception (Freire 2019).

Of how these two aspects interrelate in plasmas, Bohm initially has a fragmentary, semi-quantitative, and in fact chiefly intuitive, understanding acquired from his hands-on wartime experience, and the scant theoretical literature on the subject. Observing, for instance, the "unexpectedly rapid rate" at which electrons reach thermal equilibrium after the plasma was bombarded with a beam of fast electrons, he had understood that a *parallel* mechanism was at work beside the standard particle diffusion as derived in the kinetic theory of gases, which assumed uncorrelated particles colliding *randomly*.[37] That mechanism, able to "abstract energy from electrons

---

[36] Plasma is the state of matter produced when subjecting a gas to a temperature that is high enough to rip the electrons from their atomic shells. The result is a gas of positive and negative charges interacting via long-range Coulomb forces which, in virtue of its relative mathematical tractability would become the prototypical model of interacting many-body system—a sort of hydrogen atom for many-body physics.

[37] In fact, the necessity of modifying the standard kinetic theory to suit inter-particle forces had been already expressed, in the Soviet Union, by Anatoly Vlasov, who had reformulated in 1938 (Vlasov 1938) (published in English only in 1945) the Boltzmann kinetic equation to take account

of a wide range of velocities" particularly effectively, was identified with the excitations, and subsequent damping, of collective waves (Bohm 1947). The causes of the collective waves were the long-range electrical forces *interlocking* each particle with all the others and *correlating* their velocities, to the point that, as Bohm and Gross will put it, "a plasma displays certain forms of ordered behaviour which make a description of the system regarded as a whole more appropriate than one in which the individual particles are treated separately" (Bohm and Gross 1949). Hence, the two physicists reflect, as much as the forces induce correlations and thus complicate the treatment in the single-particle sense, in another sense they simplify it as "the system responds more or less as a unit", and affords an approximation in which "one does not have to take into account the details of individual particle motions". If, on the one hand, this presents a generic similarity with the liquid phase, on the other hand, the type of *ordering* exhibited by the plasma is only long-range. In the words of Bohm and Gross, unlike the liquid, where "the velocity of each particle is interlocked with the local average [...] the motion of a plasma shows only long range organization, while locally it is almost indistinguishable from a perfect gas."

The mathematical definition of that "local", and thus the characterisation of the length scale below which the collective ordering vanishes, is one of the objectives of the work that Bohm conducts in Princeton with his first student Gross between '46 and '48. In their analysis, this length scale is found to be approximately equivalent with the "screening (or Debye) length". Derived by Peter Debye and Erich Hückel already in 1923 in the context of electrolytes, this length, λ_D, quantified the thickness of the cloud of positive charges that surrounds each negative ion in an electrolyte. Due to this cloud, the bare ion's Coulomb potential ($\propto 1/r$) results exponentially screened ($\propto (1/r) e^{-r/\lambda_D}$). In Bohm and Gross's theory of plasma oscillations, the Debye screening length is shown to coincide with the minimum possible wavelength of the plasma oscillations, below which the collective approximation breaks down, and is put in relation to the characteristic frequency of the plasma oscillations (termed "plasma frequency"). This and other results make up the classical and complete hydrodynamical description of plasmas given by Bohm and Gross. However, it is the microscopic, first-principle derivation of the screened Coulomb potential and of the distinction between individual and collective motions that will constitute the real novelty of the collective description.

As we mention at the beginning of the section, the aim of the microscopic description of plasmas which Bohm undertakes with Pines at some point in 1947 is to apply the plasma theory to metals. In particular, one of Bohm's main motivations to do so is to explain why in metals, where the electron density is high, the effects induced by the Coulomb interactions between electrons seemed much smaller than the one that had been calculated on the basis of the single-particle approximation (see, e.g., Bardeen 1936a). As Bohm will later recollect (Hoddeson 1999):

---

of the long-range nature of the inter-particle interaction. Vlasov's equation nevertheless required an approximation to be applicable to relevant physical situations, and that is where Bohm's approach enters.

"[...] one had to understand why the effect wasn't bigger, you see. And it was basically because of screening that it wasn't bigger. And why the mobility was what it was, and so on. And so one got the idea from Debye-Hückel of a screening cloud around each particle. But the thing had to be extended to the idea of screening which moved with the particle [...]"

To unexpectedly indicate Bohm the way to the solution of this issue, and the formal tools to carry out the sought-for microscopic description of plasmas, are the recent developments of quantum electrodynamics that are being discussed in the first post-war small-scale gatherings on the Shelter Island (April 1947) and in the Pocono Mountains (March 1948). As Bohm will later recount, during the second of such meetings a viable pathway appears to him while Julian Schwinger, in an eight-hour marathon unsurprisingly legendary for its length, deftly expounds his renormalised electrodynamics. Independently from, but very similarly to Tomonaga, Schwinger had developed a relativistic version of the canonical transformation of Pauli and Fierz which—we remind the reader—allowed for going over to a new representation in which the "free" electron corresponds to the bare electron *along with* its surrounding cloud of virtual photons and electron-positron pairs. The effective result of this cloud, Schwinger demonstrates, is that the infinite mass and charge of the bare electron are reduced (renormalised) to their finite observed values.

It is the screening, renormalising, effect of the cloud of virtual electron-positron pairs on the electron's charge that strikes Bohm as analogous to the screening induced on the electrons of plasmas and metals by the surrounding charges. It is important to remark that in such an analogy, there were two significant disanalogical aspects. One was the fact that the electron-positron pairs in the vacuum are *virtual* while the surrounding charges in the real media are *real*. The second was that, differently than the single charges in the real media, the electron-positron pairs are dipoles, and thus have no net charge. For this reason, as will become relevant later, while in a plasma-like medium the *range* of the interparticle Coulomb force is reduced, in the vacuum of quantum electrodynamics—like in a dielectric medium—only the *intensity / coupling* constant of that force was reduced.

These disanalogical aspects notwithstanding, the analogy suggests Bohm to use Schwinger's mathematical method of canonical transformation to microscopically model the screening cloud moving along with the electrons in plasmas and metals. Starting now from the complete Hamiltonian for a collection of interacting electrons—featuring the kinetic energies of *all* the particles and the inter-particle electromagnetic interactions—and applying to it a series of canonical transformations à la Schwinger, Bohm and Pines are able to split it in two independent parts.[38] One part describes the long-range plasma (longitudinal) oscillation of the medium as a whole. Like Tomonaga, Bohm and Pines find that these collective oscillations have a gapped spectrum with the characteristic "plasma frequency" at long wavelengths; and, when quantised, yield a massive boson (later called "plasmon"). The other part describes the motion of the individual electrons insofar as it deviates from the mean collective behaviour. This term features the kinetic energies of the electrons

---

[38] The procedure is first carried out classically (Pines and Bohm 1950), and then quantum mechanically (Bohm and Pines 1950, 1951).

and an effective short-range Coulomb potential which corresponds to the screened Coulomb force obtained by Debye and Hückel. In this way, the new so-termed collective description of the interacting electron gas leads naturally to the concept of an effective screened Coulomb force and explains, without any ad hoc hypothesis, that experimental behaviour of plasma and metals that could not be explained with the single-particle approximation.

In the next Chapter we will see how elements from the collective descriptions of Tomonaga and Bohm & Pines, along with elements from the renormalised quantum electrodynamics, and the Hartree-Fock method are combined by Nambu into a general field theoretical framework for nuclei and solids that will act as a stepping stone for spontaneous symmetry breaking.

References

Amaldi, U. (2004). Nuclear physics from the nineteen thirties to the present day. In *Enrico Fermi: The life and legacy* (pp. 151–176). Springer.

Anderson, C. D., & Neddermeyer, S. H. (1936). Cloud chamber observations of cosmic rays at 4300 m elevation and near sea-level. *Physical Review, 50*(4), 263.

Ashrafi, B. (2004). Interview of Yoichiro Nambu by Babak Ashrafi on 2004 July 16. www.aip.org/history-programs/niels-bohr-library/oral-histories/30538.

Bardeen, J. (1936a). Proceedings of the American physical society–Minutes of the New York meeting, October 29–31, 1936. *Physical Review 50*, 1093–1101.

Bardeen, J. (1936b). Theory of the work function. ii. the surface double layer. *Physical Review 49*(9), 653.

Bethe, H., & Bacher, R. (1937). Nuclear physics a: Stationary states of nuclei. *Nuclear Physics B: Nuclear Dynamics, Theoretical, Reviews of Modern Physics, 9*, 69.

Bird, K., & Sherwin, M. (2007). *American prometheus: The triumph and tragedy of J. Robert Oppenheimer*. Vintage: Knopf Doubleday Publishing Group.

Bloch, F. (1933). Bremsvermögen von atomen mit mehreren elektronen. *Zeitschrift für Physik, 81*(5), 363–376.

Bloch, F. (1934). Inkohärente Röntgenstreuung und Dichteschwankungen eines entarteten Fermigases. *Helvetica Physica Acta, 7*(4), 385–405.

Blum, A. (2019). *Heisenberg's 1958 weltformel and the roots of post-empirical physics*. SpringerBriefs in History of Science and Technology. Springer International Publishing.

Blum, A. S., & Joas, C. (2016). From dressed electrons to quasiparticles: The emergence of emergent entities in quantum field theory. *Studies in History and Philosophy of Science Part B: Studies in History and Philosophy of Modern Physics, 53*, 1–8.

Bohm, D. (1947). Excitation of plasma oscillations. In *Abstracts of Declassified Documents*, United States. Atomic Energy Commission. MDDC. U.S. Atomic Energy Commission.

Bohm, D., & Gross, E. P. (1949). Theory of plasma oscillations: A origin of medium-like behavior. *Physical Review, 75*, 1851–1864.

Bohm, D., & Pines, D. (1950). Screening of electronic interactions in a metal. *Physical Review, 80*(5), 903.

Bohm, D., & Pines, D. (1951). A collective description of electron interactions. i. magnetic interactions. *Physical Review 82*(5), 625.

Bohr, N. (1936). Neutron capture and nuclear constitution. *Nature, 137*(3461), 344–348.

Bohr, N., & Kalckar, F. (1937). On the transmutation of atomic nuclei by impact of material particles. det kgl. danske videnskabernes selskab. *Mathematisk-fysiske Meddelelser. 14*, 10.

Bopp, F. (1948). Feldmechanische Begründung der Diracschen Wellengleichung. *Zeitschrift für Naturforschung A, 3*(8–11), 564–573.

Brown, L. M., Fujimoto, Y., Konuma, M., Maki, Z., Tsuji, T., Ishii, C., Takeuchi, M., Tamaki, H., & Tomonaga, S. (1991). Nuclear research at Riken. *Progress of Theoretical Physics Supplement, 105,* 55–68.

Darrigol, O. (1988). Elements of a scientific biography of Tomonaga Sin-Itiro. *Historia Scientiarum: International Journal of the History of Science Society of Japan, 35,* 1–29.

Dyson, F. (2018). *Maker of patterns: An autobiography through letters.* Liveright.

Fermi, E., & Yang, C.-N. (1949). Are mesons elementary particles? *Physical Review, 76*(12), 1739.

Freire, O. (2019). *David Bohm: A life dedicated to understanding the quantum world.* Springer Biographies: Springer International Publishing.

Gross, E. P. (1991). Collective variables in elementary quantum mechanics. In *Quantum Implications: Essays in honour of David Bohm.* Routledge.

Hartree, D. R. (1928). The wave mechanics of an atom with a non-coulomb central field. part i. theory and methods. *Mathematical Proceedings of the Cambridge Philosophical Society 24*(1), 89–110.

Hayakawa, S. (1991). Sin-Itiro Tomonaga and his contributions to quantum electrodynamics and high energy physics. *Progress of Theoretical Physics Supplement, 105,* 157–167.

Heisenberg, W. (1939). Zur theorie der explosionsartigen schauer in der kosmischen strahlung. ii. In *Original Scientific Papers/Wissenschaftliche Originalarbeiten* (pp. 337–362). Berlin, Heidelberg: Springer Berlin Heidelberg

Heisenberg, W. (1950). Zur Quantentheorie der Elementarteilchen. *Zeitschrift für Naturforschung A, 5*(5), 251–259.

Hoddeson, L. (1999). Interview of David Bohm by Lillian Hoddeson on 1981 May 8. www.aip.org/history-programs/niels-bohr-library/oral-histories/4513.

Hoddeson, L., Schubert, H., Heims, S., & Baym, G. (1992). Collective phenomena. In *Out of the crystal maze: Chapters from the history of solid state physics.* Oxford University Press.

Ito, K. (2002). *Making sense of ryoshiron (Quantum Theory): Introduction of quantum mechanics into Japan, 1920–1940.* Ph.D. thesis, Harvard University.

Iwata, G., Ono, K., Nambu, Y., Koba, Z., Hayakawa, S., Miyamoto, Y., et al. (1948). Abstracts of the papers presented to the symposium on the theory of elementary particles at physics department, Kyoto University, November 24–25, 1947. *Progress of Theoretical Physics, 3*(2), 207–218.

Kanamori, O. (2016). *Essays on the history of scientific thought in modern Japan.* Japan Publishing Industry Foundation for Culture.

Kanesawa, S., & Tomonaga, S.-I. (1948). On a relativistically invariant formulation of the quantum theory of wave fields. IV: Case of interacting electromagnetic and meson fields. *Progress of Theoretical Physics, 3*(1), 1–13.

Kawabe, R. (1991). From meson theory to nonlocal field theory: Hideki Yukawa in his "Days of Doubt." *Progress of Theoretical Physics Supplement, 105,* 289–293.

Kim, D. (2007). *Yoshio Nishina: Father of modern physics in Japan.* CRC Press.

Kinoshita, T. (1991). Personal recollections, 1944–1952. *Progress of Theoretical Physics Supplement, 105,* 116–119.

Koba, Z., & Tomonaga, S. (1947). Application of the "Self-Consistent" subtraction method to the elastic scattering of an electron. *Progress of Theoretical Physics, 2*(4), 218–218.

Konagaya, D. (2020). Heisenberg's visit to Kyoto in 1929 and its stimulation of young Japanese physicists. *Historia Scientiarum, 29*(3), 280–299.

Lamb, W. E., & Retherford, R. C. (1947). Fine structure of the hydrogen atom by a microwave method. *Physical Review, 72*(3), 241.

Leitch, A. (2015). *A princeton companion.* Princeton Legacy Library: Princeton University Press.

Maki, Z. (1991). Tomonaga and the meson theory. *Progress of Theoretical Physics Supplement, 105,* 149–156.

Matthews, P., & Salam, A. (1951). The renormalization of meson theories. *Reviews of Modern Physics, 23*(4), 311.

Miyajima, T. (1976a). Preface. In T. Miyajima (Ed.), *Scientific papers of Tomonaga* (Vol. 1). Misuzu Shobo Publishing Company.

Miyajima, T. (1976b). Preface. In T. Miyajima (Ed.), *Scientific papers of Tomonaga* (Vol. 2). Misuzu Shobo Publishing Company.

Miyajima, T., & Tomonaga, S. (1976). On the mesotron theory of nuclear forces. In T. Miyajima (Ed.), *Scientific papers of Tomonaga* (Vol. 1, pp. 226–261). Misuzu Shobo Publishing Company.

Nakane, M. (2019). Yoshikatsu Sugiura's contribution to the development of quantum physics in Japan. *Berichte zur Wissenschaftsgeschichte, 42*(4), 338–356.

Nambu, Y. (1948). On the relativistic formulation of the perturbation theory. *Progress of Theoretical Physics, 3*(4), 444–445.

Nambu, Y. (1949). The level shift and the anomalous magnetic moment of the electron. *Progress of Theoretical Physics, 4*(1), 82–94.

Nambu, Y. (1991). Summary of personal recollections of the Tokyo group. *Progress of Theoretical Physics Supplement, 105,* 111–115.

Nogami, M., Sato, I., Ozaki, S., Noma, S., Tanikawa, Y., Hirano, M., et al. (1947). Abstracts of the papers presented to the symposium on the theory of elementary particles at physics department, Kyoto imperial university, November 21–23, 1946. *Progress of Theoretical Physics, 2*(2), 92–100.

Oppenheimer, J., & Serber, R. (1937). Note on the nature of cosmic-ray particles. *Physical Review, 51*(12), 1113.

Oppenheimer, R. (1949). Correspondence from Robert Oppenheimer to general Douglas MacArthur, 20 February, Director's office: Member files: Box 139: Tomonaga, Sin-itiro, Shelby White and Leon Levy Archives Center, Institute for Advanced Study, Princeton, NJ, USA.

Oppenheimer, R. (1954). Director's report transcripts, drafts, board of trustees records: Board-general: Box 1: 1954. Shelby White and Leon Levy Archives Center, Institute for Advanced Study, Princeton, NJ, USA

Pines, D., & Bohm, D. (1950). Proceedings of the American physical society–Minutes of the meeting at Washington, April 27–29, 1950.

Proceedings of the Japan-USA collaborative workshop (1991). Appendix D. Table: Record of "Meson Meetings". *Progress of Theoretical Physics Supplement 105,* 50–51.

Rohrlich, F. (1950). Quantum electrodynamics of charged particles without spin. *Physical Review, 80*(4), 666.

Sakata, S. (1954). Proceedings, international conference on theoretical physics: Kyoto and Tokyo, September 14–24, 1953.

Schweber, S. (2020). *QED and the men who made it: Dyson, Feynman, Schwinger, and Tomonaga.* Princeton series in physics. Princeton University Press.

Takabayasi, T. (1991). Nonlocal theories and related topics. *Progress of Theoretical Physics Supplement, 105,* 270–286.

Taketani, M. (1971). Conflict between matter and field: An analysis of the difficulties of the theory of elementary particles. *Progress of Theoretical Physics Supplement, 50,* 89–97.

Tati, T., & Tomonaga, S.-I. (1948). A self-consistent subtraction method in the quantum field theory. I. *Progress of Theoretical Physics, 3*(4), 391–406.

Tomonaga, S. (1938). Innere Reibung und Wärmeleitfähigkeit der Kernmaterie. *Zeitschrift für Physik, 110*(9–10), 573–604.

Tomonaga, S. (1946). On a relativistically invariant formulation of the quantum theory of wave fields. *Progress of Theoretical Physics, 1*(2), 27–42.

Tomonaga, S. (1947). On the effect of the field reactions on the interaction of mesotrons and nuclear particles. iii. *Progress of Theoretical Physics 2*(1), 6–24.

Tomonaga, S. (1955a). Introduction: Progress in meson theory in Japan. *Progress of Theoretical Physics Supplement, 1,* 1–6.

Tomonaga, S. (1955b). Zur theorie des Mesotrons. i. *Progress of Theoretical Physics Supplement, 2,* 1–20.

Tomonaga, S. (1966). Development of quantum electrodynamics, personal recollections. https://www.nobelprize.org/prizes/physics/1965/tomonaga/lecture/.

Tomonaga, S. (1968). Reminiscences. *Progress of Theoretical Physics Supplement, E68,* 3–5.

Tomonaga, S. (1976). Bemerkung Über die Streuung der Mesonen am Kernteilchen. In T. Miyajima (Ed.), *Scientific papers of Tomonaga* (Vol. 1, pp. 214–225). Misuzu Shobo Publishing Company.

Tomonaga, S. (1976b). The development of elementary particle theory—Discussions centring on the divergence difficulties. In T. Miyajima (Ed.), *Scientific papers of Tomonaga* (Vol. 1). Misuzu Shobo Publishing Company.

Tomonaga, S. (1976c). Miscellaneous: Letters from America. In T. Miyajima (Ed.), *Scientific papers of Tomonaga* (Vol. 2). Misuzu Shobo Publishing Company.

Tomonaga, S., & Miyajima, T. (1976). *Scientific papers of Tomonaga* (Vol. 1). Scientific Papers of Tomonaga: Misuzu Shobo Publishing Company.

Tomonaga, S., & Miyajima, T. (1976). *Scientific papers of Tomonaga* (Vol. 2). Scientific Papers of Tomonaga: Misuzu Shobo Publishing Company.

Tomonaga, S., & Oppenheimer, J. R. (1948). On infinite field reactions in quantum field theory. *Physical Review, 74*(2), 224.

Tomonaga, S.-I. (1950). Remarks on Bloch's method of sound waves applied to many-fermion problems. *Progress of Theoretical Physics, 5*(4), 544–569.

Van Hove, L. (1949). Relativistic terms in the interaction between nucleons in pseudoscalar and vector meson theory. *Physical Review, 75,* 1519–1523.

Vlasov, A. (1938). On high-frequency properties of electron gas. *Journal of Experimental and Theoretical Physics, 8*(3), 291–318.

Watson, K. M., & Lepore, J. V. (1949). Radiative corrections to nuclear forces in the pseudoscalar meson theory. *Physical Review, 76,* 1157–1163.

Weart, S. (1992). The solid community. In *Out of the crystal maze: Chapters from the history of solid state physics.* Oxford University Press.

Wentzel, G. (1940). Zum Problem des statischen Mesonfeldes. *Helvetica Physica Acta, 13*(4), 269–308.

Yukawa, H. (1950). Quantum theory of non-local fields: Part i. Free fields. *Physical Review, 77,* 219–226.

Yukawa, H. (1991). Appendix E. The path i have followed: At Shimogamo, Kyoto, in April 1974. *Progress of Theoretical Physics Supplement, 105,* 52–54.

Yukawa, H., & Sakata, S. (1939). Mass and mean life-time of the meson. *Nature, 143*(3627), 761–762.

Chapter 3
Yoichiro Nambu and the Collective Description of Many-particle Systems

In this chapter we retrace the salient steps of Nambu's eclectic education as a physicist and his resolute quest for a rigorous non-perturbative treatment of the many-body problem, primarily motivated by the problem of the nuclear forces but inspired by methods of solid-state physics. Instrumental to this quest are Dirac's material interpretation of the vacuum, Nambu's early exposure to busseiron physics—a mixture of statistical physics and solid-state physics characteristic of Japan—as well as to Tomonaga's approach to renormalisation and the collective description, whose lessons Nambu generalises and synthesises in one powerful framework. Despite the lack of success in the intended realm of nuclear phenomena, this framework will later enable Nambu to successfully formalise superconductivity in the language of field theory, as we will detail in the next chapter.

3.1 *Lehrjahre* and Inclinations of the Young Nambu

Initially not one of Tomonaga's students, Yoichiro Nambu enters Tokyo Imperial University in 1940 to study under the formal supervision of the nuclear and *busseiron* physicist[1] Kiichiro Ochiai—who had spent a period under Werner Heisenberg in the 1930s—and graduates in two and a half years, his course of study being cut short by Japan's intervention in World War II. Alongside the standard courses of the physics

---

[1] As we will expand later, "busseiron kenkyu", translatable as "research on the theory of 'physical' matter" (the composition *bus-sei-ron* can be rendered as matter-properties-theory), is a peculiarly Japanese disciplinary area of physics, sitting in between condensed matter physics and statistical physics, quite mathematically laden and specifically oriented to theoretical approaches to the properties of matter (private communication with Hiroto Kono). For more details on the tradition and the historical characters of *busseiron* physics see Kono (2020).

© The Author(s), under exclusive license to Springer Nature Switzerland AG 2022   47
R. Gaudenzi, *Historical Roots of Spontaneous Symmetry Breaking*,
SpringerBriefs in History of Science and Technology,
https://doi.org/10.1007/978-3-030-99895-0_3

curriculum, Ochiai's students gather in self-organised weekly seminars to read and report on more timely and advanced topics mainly using three reference texts (Nambu 1991): Ralph Fowler's "Statistical Mechanics", Walther Heitler's "Quantum theory of radiation", and the first voluminous review on "Nuclear Physics" authored by Hans Bethe and Robert Bacher a few years before. Those students of the group, including Nambu, who are inclined to what was then called "high-energy nuclear physics" are, however, not fully satisfied under Ochiai's supervision and the environment of Tokyo University (Nambu 1991). In effect, Ochiai, like other supervisors at Tokyo Imperial University, apparently tends to discourage the students from undertaking the more "daring" particle physics, in favour of nuclear physics or condensed matter physics, "safer" from the standpoint of a prospective career and applications.[2] If the Imperial University was among the strongholds of these latter sub-fields, the Institute for Chemical and Physical Research (*Riken*)—as we had discussed in Sect. 2.1.3—was home to a very active group of experimental and theoretical particle physicists headed by Nishina and Tomonaga. Participation in just a couple of the weekly seminars led by them is enough to leave a profound mark on Nambu and a few other young and new attendees both for the topics discussed and Tomonaga's qualities as a teacher, an impression which strikes them as the burgeoning research they were looking for (Ashrafi 2004).

Upon his return to Tokyo at the beginning of '46 from the war station where he had been relocated, Nambu finds out that Tomonaga had already assembled a group of young theoreticians from among the ones available. These were the few students that either were not called to the army or who, unlike Nambu, were stationed close to Tokyo and could thus more easily commute to the city to attend Tomonaga's Sunday courses (Hayakawa 1991). Remaining for the year to come in the condition of a *shugyōsha*,[3] with minimal academic duties and literally living in his office due to the extreme poverty of post-war period (Brown and Nambu 1998), Nambu pursues his core curiosity in particle physics by watching his office-mate and Tomonaga's right hand Ziro Koba at work. Through him, he "learns what the Tomonaga group is doing" (Nambu 1991), and meanwhile carries out independent work on the formal aspects of second quantisation. Owing to the post-war paucity of means, many of Nambu's peers, too, spend their nights and days in the shabby office rooms of the Imperial University. The closed universe of forced proximity and the extraordinary situation call for an equally extraordinary immersion and density of intellectual exchange.

---

[2] In an interview (Ashrafi 2004), to the question of whether people at that time distinguished between particle and nuclear physics, Nambu answers "No, not yet. But nuclear physics was what you learned". This expression concisely captures the view of subatomic physics, at least from the Japanese perspective, as segmented between nuclear physics as a rather established and commonly taught body of knowledge, and "particle" physics as whatever else pertained only to the forefront, practised exclusively in laboratories and discussed only in certain research circles (see Sect. 2.1, and also Nambu (1991); Kinoshita (1991)).

[3] In feudal Japan the term was used for the itinerant samurai who, while looking for a lord, practises and hones his skills without the protection of his family or school. The term comes to metaphorically indicate students that did not have an official supervisor when, a few years later, Nambu and his colleagues would promote an informal program of exchange of these students and called it accordingly *musha shugyō* (Ashrafi 2004).

Years later, questioned on how much time he was able to devote to physics back then, Nambu will answer that, with the exception of time spent out looking for food, he studied "day and night, actually," recalling also that "there was another guy who lived in my room, and still another in the adjacent room and day and night we talked physics" (Ashrafi 2004). One of the two, Giichi Iwata, senior to Nambu, is an eccentric physicist, and erudite in European literature, Ancient Greek and Latin. Iwata, Nambu recollects, teaches him "all those things besides physics" and introduces him to various topics in modern physics, from nuclear theory to statistical mechanics. One of these is the Ising model, which is being discussed also right next door where the numerous solid-state physics group led by Ryogo Kubo is quartered, in line with the main vocation of Tokyo University. Kubo was among the few who had managed to avoid army recruitment, and had kept working steadily during wartime with a restricted group of students (Hoddeson et al. 1992).

Much discussed among solid-state physicists in those years, and known in Japan with a delay of a few years, is the exact solution of the two-dimensional Ising model that Lars Onsager had provided via a complicated mathematical procedure. Onsager's solution represents the first theoretical model of a phase transition from a disordered to an ordered (magnetic) configuration of matter. In particular, it demonstrates the existence of a such a transition in a two-dimensional lattice of spins interacting via simple short-range forces (Onsager 1944). Focused on the mathematical aspects of Onsager's procedure and enticed by the challenge of simplifying it,[4] as his first work as a graduate, Nambu ends up providing an alternative algebraic formulation of it in the spring of 1947 (published later in Nambu (1950a)). By means of a "striking analogy" between the eigenvalue problem of ordinary quantum mechanics and that identified by Eliott Montroll and Onsager as yielding the partition function of a Ising ferromagnetic system, Nambu manages to reformulate their solution by borrowing a conceptual tool from field theory. In analogy to the procedure of second quantization, the alternative route he finds is based on an algorithm through which the spin operators featuring in the Ising Hamiltonian are regarded in turn as the elements of a space on which other "*super*-operators" act. Reinterpreted in this way as a sort of "third quantisation", from a higher configuration space, the Ising Hamiltonian takes the form of a rotation operator, the eigenvalues of which can be readily found.

While this alternative algebraic method is being worked out in detail, the discovery of π-meson and Lamb shift in early 1947 hurriedly brings Nambu's focus back to particle physics subjects, to involve him with the ongoing activity of Tomonaga's group, which was by then feverishly occupied with squeezing numerical predictions out of the so-called "self-consistent subtraction method" (see Sect. 2.1.3). Participating in the seminars on a regular basis but perhaps unwilling to join the sheer computational work, Nambu appears to have maintained a freedom of action and a certain degree of independence from these activities. Following the suggestion of

---

[4] On a few occasions, when recollecting those tough years, Nambu underlines his fascination for such problems. In one such recollection (Nambu 1995), he says: "This problem [finding the ground state of an Ising-like magnetic system] cast a spell on me, and after a while I found a simple algebraic derivation of the solution".

Tomonaga, he concentrates on an *alternative* pathway to calculate the Lamb shift and the correction to the electron's magnetic moment (Nambu 1949a; Tomonaga 1952).[5] In such independence, Nambu's previous reflections on the Ising model, still unpublished and seemingly far removed, find a place in the renormalised quantum electrodynamics developed by Tomonaga. In three papers published in 1949, Nambu applies an analogue of the algebraic "method of third quantisation" that he had developed for the Ising model to obtain a tidy mathematical reformulation of the as yet open problem of normal ordering in quantum electrodynamics (see Nambu 1949b, and references there in).[6]

Spurred by the challenge of mathematically reformulating and translating given physical problems into other mathematical forms, these early works betray a playful, conceptually thoughtful, focus on mathematical *methods* in themselves as the one pivotal aspect unifying different disciplinary subjects across *domains* of applications, a focus that will be the distinctive feature of the whole of Nambu's future production. Based on deft, creative and original mathematical reformulations, Nambu's works at this stage are, however, not yet daring enough to touch on the substance of the underlying physics to which they apply. As Tomonaga presciently says, in tracing the profile of Nambu to Oppenheimer (Tomonaga 1952):

> Yoichi [sic.] Nambu [...] had been a participant of my seminar for theoretical physics since 1947.[7] [...] The characteristics of his works is, in a word, an endeavour to look into problems from a perspective somewhat different from the orthodox one. At present such an endeavour has not always been very successful, sometimes resulting only in too much mathematical artifice without giving something physical; still I am glad to have a physicist of this type looking forward to his future achievement. He himself seems to be aware of this weakness and is making an effort to carry out works that have more physical contents [...].

Indeed, Nambu's distinctive style of doing physics will emerge when the focus on the mathematical methods and his prowess in, and passion for, mathematical frameworks will more closely engage with "physical contents"—to take up Tomonaga's words. As we will see, this style will find its best expression in that integration, translation, and generalisation of existing ideas and methods to&from high-energy physics and low-energy physics that characterise at its essence the pathway to spontaneous symmetry breaking. The result of a maturation in the vision of physical problems

---

[5] It is not by chance that, in contrast to most of the works of Tomonaga's core group, Nambu's results are reported in a single-author paper issued in parallel to the joint works of Tomonaga's group.

[6] At the unfinished stage of development in which quantum electrodynamics finds itself in 1949, it is not yet clear how to systematically handle the calculations for higher-than-second order processes. The reason was that such processes are written as a long product of field operators which implicitly contain a number of distinct elementary processes—corresponding to the individual Feynman diagrams, in Feynman's formulation. A mathematical tool is therefore needed to "extract" each meaningfully ordered configuration by expanding the long product into a sum of pairs of operators. This tool in its general form will be provided by Gian-Carlo Wick the following year.

[7] As documented, in fact Nambu had participated in the seminar even before, when still a student, but somewhat more unofficially.

and of the freedom allowed by a new environment (Nambu 2016), the more direct engagement with the physical aspects that Tomonaga expected begins to manifest with the transfer to the young Osaka City University.

3.1.1 Nambu's Nuclear Program and the Root of Spontaneous Symmetry Breaking

In Osaka, Nambu is appointed professor on the recommendation of Tomonaga. His appointment is part of the plan promoted by Yuzuru Watase—then the department chairman—of opening the physics department of the recently founded university with a group of young and outstanding physicists (Low 2005). Being the first professor to join the department, as Nambu recalls, "there were no senior professors to defer to" and "no need for formal lectures as there were only two students" (Nambu 2016). Not restricted to the academic sphere, the anecdotes and the tone of Nambu's recollection of the Osaka period transmit the light-heartedness, the "easy life there at that time", the relative relief after the hardship of the immediate post-war years in Tokyo which, differently than the Kyoto area, had been heavily bombed. As the head of a small group of enthusiastic émigrés from Tokyo University[8]—all promoted from one day to the other from mere assistants at best, to higher roles according to their seniority— in Osaka Nambu is now, therefore, truly free to pursue his own research questions, setting up a collaboration with each of them.

Testifying to the inviting atmosphere and to the wish of challenging the traditional "feudal" borders between the Japanese universities, the group establishes an informal inter-university exchange, symbolically appealing to the ancient Japanese practice of *mushashugyo* (see Oneda 1991; Ashrafi 2004; Nambu 1995, and Footnote 3). One after the other, some young students from Tokyo are invited for short stays in the group to promote the interactions with the physicists of the group, and at the same time to maintain high spirits and the group dynamic. Free from matters of priority, preliminary sketches and emerging half-baked ideas are quickly and informally presented through the handwritten tabloid journal *Soryushiron Kenkyu* (Particle Theory Research).[9] Completely unknown outside of the Japanese circle of physicists, some of these communications uncover heuristic steps which would never reach the surface of published material, and have been an important resource for the conceptual reconstruction we present.

---

[8] Satio Hayakawa who had worked under Tomonaga, Yoshio Yamaguchi probably a student of Tomonaga too, and Kazuhiko Nishijima. A fourth one, Tadao Nakano, joined in the spring of 1951.

[9] The journal, which Nambu defines on a few occasions as a tabloid magazine and "one of my favourite journals" (Nambu 1995), had been founded in October 1948 by Seitaro Nakamura with the aim of reporting on on-going research in brief, *hand-written* and informal communications, including correspondence, translations in Japanese of international papers, etc.

A representative sample of the versatility and breadth of some of the by-then young Japanese "particle" physicists—and especially those who had been part of Tomonaga's study group—Nambu's collaborators have, like him, invariably been exposed to, and are acquainted with, a variety of problems and techniques from low-energy physics and high-energy physics. If among the physicists of the previous generation (e.g., Heisenberg, Fermi, Wigner) such a familiarity was relatively common, in Europe and America this becomes rarer already among those born after 1905, under the effect of the increasing specialisation between sub-disciplines of physics (Weart 1992). This is not the case in a hitherto isolated Japan, where resources are scant and physicists still have to be educated and learn as broadly as possible because, among other reasons, they do know what they will end up doing in their scientific careers (see also Low et al. 1999).[10] The research activity of Nambu's group as well as the future production of its members evidence this breath, versatility, and diversity of interests. It rambles from cosmic rays physics to nuclear reactions in stars to particle physics with a blend of methods which make them constantly walk across the boundary between low-energy statistical physics and high-energy physics, and features a mixture of fundamental and phenomenological approaches.

With a similar blend of methods and the different angles that these methods afforded, Nambu will also attack the problem of the nuclear forces, aiming at a mathematically adequate formulation of Yukawa meson theory (in the sense specified below) and the construction of a nuclear potential. As it shines through from the first page of the notebook entitled *Dichtung und Wahrheit*,[11] this becomes in effect Nambu's central and at the same time chimeric research target from the beginning of the Osaka period, it prefigures the major motive of struggle of the two-year period in Princeton, and can be regarded in retrospect as the motive force leading to spontaneous symmetry breaking. Through the ideas and attempts appearing therein, the notebook evidences the resolution to surpass the present, stalling situation of nuclear physics by tackling the as yet unresolved problems of the nuclear forces from first principles. As the opening note dated 21st January '50—the day of Nambu's 29th birthday—shows (Nambu 1950c), this would be achieved through a "New Formalism" for the "relativistic many-body problem" by means of which one could rigorously derive the properties of the nuclei and of the nuclear forces. In particular, the targeted properties were the so-called saturation of nuclear forces, the anomalous magnetic moment of the nuclei, the presence of "magic numbers"—nucleons numbers for which the nuclei are particularly stable—and, with that, the success of the shell model. What motivates Nambu in the pursuit of such a formalism is the fact that still none of the meson theories of nuclear forces based on the Yukawa hypothe-

---

[10] Some exception to this rule is seen in the school of Yukawa, based in Kyoto. The physicists educated under that school seem to be from the beginning more strictly focused on nuclear and particle physics and less mathematically inclined.

[11] Nambu uses for this notebook the title of Goethe's autobiography. This is not a chance nor an isolated case: other expressions like *lehrjahre* and *wanderjahren*, used by Goethe to describe the stages of the intellectual pathway of his "autobiographical" Wilhelm Meister appear here and there in Nambu's writings.

sis was experimentally adequate, the years of attempts notwithstanding.[12] While the hypothesis had certainly provided a qualitative picture of the nuclear interaction and managed to explain some of its properties, the theories had not hitherto been able to account for the other aforementioned equally prominent properties.

Commencing when nuclear physics faces this crisis, the quest that Nambu undertakes is somewhat anachronistic. It is a solitary quest and appears thereby as an obstinate move when most nuclear physicists are on the way to abandoning that nucleon-meson model of the nucleus—and the understanding of the nuclear interactions in nuclei in terms of nucleons exchanging mesons—that had hitherto fuelled the hopes for an explanation of the nuclear phenomena (Furlan and Gaudenzi 2021). They abandon it mostly in search of an alternative, more fundamental solution, and hunting for fresh phenomena in that sub-nuclear world of elementary particles whose rich prospects had just been opened by new instrumentation: the particle accelerators and detectors for high-energy cosmic rays.[13] Certainly not exceedingly seduced by such prospects, nor by the tempting opportunities offered to theoretical physicists by the flow of data, one ingredient behind Nambu's decision to reformulate the nuclear problem is the clear and idiosyncratic preference for, and satisfaction in, neat, large-scope mathematical frameworks and a parallel interest in many-body problems. But there is another, more direct justification. His interest in these experiments notwithstanding, Nambu's position evidences a certain scepticism as to them being the source for the solution. Spoken through the doctoral thesis of Nambu's student Kazuhiko Nishijima (1951):

> When the serious difficulty concerning the singularity of the meson potential became emphasized, and so discouraging was this difficulty, as compared with the Coulomb potential, for both theoretical and experimental reasons, that the efforts of early meson physicists were concentrated on this problem. Recently, however, the laboratory studies of mesons were rapidly advanced, and most meson physicists are interested in the mechanism of the production and capture of pi-mesons leaving the problem of "nuclear forces" untouched. Still we have reasons to believe that this phenomenon, though implicitly related to the properties of mesons, will give us some information about the correct method to be employed in the meson problem [...].

As can be perceived in these words and in continuity with the spirit of Tomonaga's work, Nambu insists on a more conservative attitude towards the matter, the focal point of which remains the search for a mathematical framework that suits the nuclear problem. Before one can decide, he writes in 1950 (Nambu 1950b), "between the existing "korrespondenzmässigen" theories and any more revolutionary ones", one

---

[12] For a review of the attempts and a picture of the crisis of nuclear physics see, e.g., Wentzel (1947); Tanikawa (1948), Fermi's textbook (Orear and Fermi 1950, p. 111) and the textbook on meson theory by Robert Marshak (Marshak 1952, p. 1). These sources are put into context in Furlan and Gaudenzi (2021).

[13] The solution of the nuclear problem will indeed be found, more than a decade later, at the more fundamental level with the quarks. It is worthwhile to mention that, in parallel with his search for a formalism for the nuclear problem, Nambu works for about half a year on the V-particle just discovered in cosmic rays (Nambu et al. 1951). Left to his two students Nakano and Nishijima, this work would lead them to the formulation of the Nakano-Nishijima law, precursor to the quark idea.

should be sure about the adequacy and the validity of the formal means through which the conclusions have been hitherto derived in the existing models.[14] In other words, before dismissing the Yukawa hypothesis, one should be sure to have drawn its conclusions rigorously, in particular, to have in the first place re-examined "the old concept of potential" in light of the recent method of quantum electrodynamics so as "to conform it more closely to the field theoretical picture" (Nambu 1950b). This is what dictates the necessity and significance of the "New Formalism" that Nambu seeks.

Having conducted the critical examination on the concept of potential, the conclusion Nambu reaches is that the sought-for formalism cannot be based on perturbative calculational tools hitherto used in quantum field theory and one must thus somehow replace them. In this way, he re-grafts that discussion of non-perturbative methods which had characterised most of Tomonaga's research into the now more solid field theoretical framework that had been given to quantum electrodynamics. Having excluded the more expectable option of using the standard tools of quantum field theory as a base for his non-perturbative formalism, the alternative option—which figures already in the plan Nambu lays out in the 1950 notebook—is to construct a "*busseiron* model" of nucleons and mesons. This is a model which would preserve the scaffolding of quantum field theory but draw from the corpus of many-body mathematical methods that had been developed in the context of the low-energy physics to address the properties of solids. While the direction is set, what to draw on exactly and how to construct such a description is not a priori clear to Nambu. The first tile of the mosaic comes about through the phenomenological treatment of a nuclear reaction.

3.2   The Concept of Apparent Vacuum: Extending Field Theory to Real Media

As is common practice in the group, in parallel with the fundamental and analytical approach to the nuclear problem, Nambu and colleagues also tinker with other ideas related to, and motivated by, a phenomenological modelling of nuclear reactions. This is carried out by making use of a mixture of tools from statistical mechanics and dispersion theory, analysing, as they write, "the mesonic phenomena in some-

---

[14] Ever since the first works on the meson field in 1935 (Yukawa 1935), Yukawa and collaborators had developed the procedure to obtain the interaction potential between a proton and a neutron and calculate the bound states in analogy with the procedure adopted in relativistic quantum mechanics to treat the electron-proton bound state—this is the sense of the "korrespondenzmässigen". In that procedure, criticised by Nambu, the potential is worked out, from a Hamiltonian analogous to that of classical electrodynamics, as the *second-order* term of the perturbation series ordinarily used in quantum theory; and the potential so obtained is then inserted, as if originating from a static point source, in the Schrödinger (or Dirac) eigenvalue equation to compute its consequences, e.g., the scattering properties for unbound systems, or the energy levels of the deuteron bound system. All computed with this same perturbative scheme, over the years a plethora of different meson interactions had been proposed. These are the existing models Nambu refers to.

what different manners as compared with the viewpoint of current meson theories" (Yamaguchi 1951). Conducted in general to gain a more quantitative grasp of the meson-nucleon forces, these analyses are, like most of the phenomenological modelling activity of these years, strongly spurred by the increasing amount of data collected by the particle accelerators. Some of the experiments use as a target complex nuclei,[15] and to obtain the sought-for quantitative match between data and calculations, the latter should include the effect of the nuclear medium in the analysis of the elementary processes, at least in an approximate fashion. To account for such an effect is also the purpose of the detailed calculations conducted by Nambu's student Yoshio Yamaguchi at the beginning of 1950 on the probability of pion capture processes in complex nuclei. In a capture process, a pion in the lowest orbit of an atom is absorbed by one of the nucleons in the nucleus, which is as a result promoted to a higher energy state with a certain probability. Even if the reaction occurs ultimately between the pion and, say, one proton, that probability—not dissimilarly from a chemical reaction—is affected by the surrounding nuclear medium. Regarding the nucleus at its simplest as a degenerate Fermi gas, the largest of such effects is due to Pauli's exclusion principle in that the pion can be absorbed only by those protons that can be kicked above the Fermi level by a high enough pion momentum.[16]

In this context, Nambu, who is thinking of ways to view the nuclear reaction from the standpoint of relativistic field theory, recognises in the problem of pion capture the possibility of another effect due to the presence of nuclear matter. "The direct motivation of what I will discuss here is Mr. Yamaguchi's detailed calculation" concedes Nambu in a handwritten communication entitled 'The pion self-energy' in *Soryushiron Kenkyu* (Nambu 1950d), but "[...] because it is theoretically interesting for its own sake, I will propose it as a problem." The problem is to estimate the effect that the bulk nuclear matter of the nucleus might have on the *virtual* processes of the incident pion, i.e., on the pion self-energy. Reasoning in simple terms, such an effect could be written as the difference between the self-energy in the free state (pion in vacuum) and that in the "bound" state (pion in the nuclear medium). If measurable, Nambu suggests, this difference could be used to experimentally test the validity of the picture of vacuum polarisation in quantum field theory as the Lamb shift had provided a test bed for mass renormalisation.

In fact, there is a further reason why Nambu is particularly keen on computing the change of the self-energy of the pion that results from its being inserted in the nuclear matter. He foresees that therein lies a way to account for the effect of the surrounding nuclei on the nucleon-nucleon force potential by giving the pion new effective propagation properties as compared to the free pion. In particular, in the nuclear matter the latter should have acquired an increased self-energy/mass $\mu(p)$ possibly proportional to the density of the nuclear matter p. The increased mass, inserted in the Yukawa potential in the place of the free pion mass, $1/r \, e^{-\mu(p)r}$,

---

[15] By "complex" we mean here in general nuclei with relatively high atomic numbers. The first such experiments involve carbon as the target nucleus. See, e.g., Lax and Feshbach (1951).

[16] This effect is the analogue of that due to the occupied Fermi sea in the process of photoelectric ionisation in, e.g., metals.

would allow one to effectively include in the force potential those many-body effects that were problematic to treat with the standard force potential approach, and could potentially explain the hitherto unexplained properties of the nuclear force.

In this prospective framework vaguely intuited by Nambu, the problem remains of how to practically compute with some degree of accuracy the effect of nuclear matter on the pion's self-energy. Quantum electrodynamics, in which only elementary scattering processes had hitherto been considered and was in that sense a "few-body" theory, could not help. As described below, to evaluate the sought-for contribution in the 1950 communication, Nambu resorts to Dirac's "hole theory". With that, Nambu reintroduces in the context of the modern field theory the picture of the vacuum as a non-interacting, locally polarisable medium used by Dirac twenty years earlier. As we will see, this reintroduction has unexpectedly significant consequences on the future of physics: it opens the way to the "substantialisation" of the vacuum of field theory, a process which will ascribe to the vacuum interactions and properties that are analogous to those of ground states of various solid media, progressively transforming that inert Fermi gas originally postulated by Dirac into a superconducting medium.

3.2.1   Nuclear Matter as a "di-mesic" Medium

In 1928, Paul Dirac had obtained a relativistic generalisation of the Schrödinger equation for the free electron. In it, beside the electron in its two spin states—the spinor that Wolfgang Pauli had previously introduced—there appeared, somewhat surprisingly, a symmetric pair of solutions but with *negative* total energy (the rest mass plus kinetic energy). Since these states could not be discarded without compromising the relativistic covariance of the theory, and yet had to be occupied "so that our ordinary electrons of positive energy cannot fall into them" (Dirac 1931), Dirac's idea was to reinterpret what we consider empty space—the vacuum in the standard sense—instead as a medium homogeneously filled with negative energy electrons.[17] In other words, the electrons that we observe would be "sustained" by a "sea" of negative-energy electrons filling the vacuum, unobservable though real enough to obey the Pauli principle.[18] In this picture, the ground state of the universe corresponds to the configuration in which all the negative-energy states are filled with electrons, and the otherwise unexplainable stability of the visible matter is rendered compatible with the new equation. Travelling through such a "dense" vacuum, for example, a photon of appropriate energy could excite one such negative electron and

---

[17] A detailed account of the evolution of Dirac's ideas can be found in (Wright 2016).

[18] As well documented in Wright (2016), the analogy to the sea appears for the first time in 1934 (we re-quote from the same reference): "We can then assume that in the world as we know it, nearly all the −ve [negative] energy states are filled, with one electron in each state, and that the vacuities or holes in the −ve energy distribution are the positrons. A rough picture is to compare it with the sea, a bubble in the sea corresponding to a positron, just as a drop of water in the air corresponds to an ordinary electron. The sea is a bottomless one, and we do not consider at all how it is supported from below, but are interested only in events near its surface".

"push it" out of the vacuum like an electron is expelled from a metal or semiconductor via photoelectric ionisation. As in the standard ionisation of a physical medium, the process of photoelectric ionisation generates in fact not only one electron, but also leaves behind one hole in the vacuum "medium".[19] This hole, Dirac had reasoned, is the result of the *removal* of negative energy and negative charge from the vacuum, and would thus be observed as the *occupation* of a particle state with positive energy and positive charge. The predicted existence and the mechanism of photo-generation of this positive-electron (also called anti-electron or positron) would indeed be confirmed just two years later, and explained those tracks mirror-symmetric to electrons that had been observed, but hitherto ignored.

With the further development of quantum electrodynamics and the conception of the electron as a field, Dirac's physical ideas of vacuum, removal from, and occupation of, were further reinterpreted as processes of creation and annihilation from a field-theoretical vacuum—with the latter being regarded at most figuratively as "filled" with particles (Weisskopf 1949). From this point of view, all the particles we observe are thought of as truly *created*, always temporarily, out of a vacuum state in which the particles are not physical but only "latent".[20] This removed the question of the stability of the matter, the need to invoke physical particles obeying the Pauli principle, and thus avoided the question of the ontological reality of the negative energy particles of the Dirac sea, and with that also its quite problematic infinite charge and energy density. In this transition from the old to the new field-theoretical framework, a clean picture of particles' self-energy emerged which avoided any reference to the Dirac sea. Instead of being visualised as polarizing the vacuum medium, for example, the travelling photon in its course simply *annihilated* in the vacuum thereby stimulating the temporary *creation* of an electron-positron pair from the vacuum, which in turn annihilated back into the photon.[21]

In general, one can say that, having served as a scaffolding, Dirac's material interpretation of the vacuum was therefore removed once the edifice of quantum electrodynamics had come to stand independently. That idea had indeed been removed, but not eradicated. With the issue about its ontology suspended, in Japan the concept of the Dirac sea seems to have survived on a parallel undercurrent course, employed as an aid to the intuition and to perform phenomenological calculations even after the emergence of full-blown electrodynamics.[22]

It is in this context that Nambu unearths and cleverly applies to the pion traversing the nuclear medium Dirac's material interpretation of the vacuum and the idea of photon polarising it. Drawing from Dirac's picture, Nambu conceives the Fermi sea of real nucleons as the positive energy sector—as a sort of observable tip—of the underlying Dirac sea of negative nucleons. In this sense, the two seas are merged

---

[19] Dirac's understanding of the nature and properties of the holes is indeed based on an analogy to a many-electron atom. See, e.g., Dirac (1930).

[20] The idea of latent particles is used, for instance, in 1949 by Viktor Weisskopf (1949).

[21] As a fossil residue of the original analogy, the process responsible for the self-energy of a photon was, and still is, referred to as "vacuum polarisation".

[22] We advance some arguments on the reasons behind this course in Chap. 5.

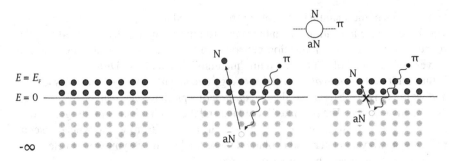

Fig. 3.1 On the left-hand side, the Fermi sea of nucleons (positive energy sector: $0 < E < E_F$) "emerges" from the Dirac sea (negative energy sector: $-\infty < E < 0$), and the two form a continuum of particles obeying the Pauli principle. On the right-hand side, the nucleon-antinucleon (N-aN) pair creation process contributing to the pion, π, self-energy is schematically represented. In the presence of the Fermi gas above the Dirac sea the processes that involve particles whose energy is smaller than the Fermi energy are forbidden by the Pauli principle. This results in a finite deviation of the pion self-energy—essentially given by the change in the integration bounds in the self-energy integral—from its value in vacuum

into one homogeneous continuum in which the nucleons of negative energy *virtually* excited by the pion to positive energy levels obey the Pauli principle exactly as the *real* nucleons do. For this reason, a part of the *virtual* processes contributing to the pion self-energy that would occur in vacuum are forbidden in the nuclear medium due to the presence of the real nucleons (see Fig. 3.1). This results in an observable deviation of the pion self-energy that can be easily estimated. In Nambu's words (Nambu 1950d)[23]:

> Because nucleons in a nucleus occupy not only negative energy but also a finite positive energy E_F [the Fermi energy/momentum], the vertical (*sic*) nuclear pair [the nucleon-antinucleon pair virtually generated by the pion] appears differently than in a vacuum. This will give a finite deviation of the self-energy. [...] because the pion [in its orbit around the nucleus] can be supposed to be at almost complete rest, nucleons of positive energy cannot immediately absorb it by Pauli's exclusion principle. In a similar way, nucleons of negative energy cannot become nucleons of finite positive energy whose momentum is less than E_F by the absorption [...] of the pion. The latter is a deviation from the case of a vacuum, which can be immediately estimated with a simple perturbation calculation.

As in Dirac's considerations, the matter-like aspect of the vacuum is based on the analogy between real and virtual particles as far as the Pauli principle goes. Such aspect is, however, invoked for a different purpose and an opposite rationale. The purpose is not to explain how the real fermions are rendered stable by the presence of the underlying Dirac sea, but to estimate how the presence of the bulk of real fermions (nucleons, in this specific case) modifies the virtual transitions from the Dirac sea. Nambu's rationale is thus not that of saving the real matter preventing the downward transitions to the negative states, but to show how the Fermi gas prevents a part of the upward transitions to the positive states.

---

[23] The translation from Japanese kindly has been kindly provided to us by Dr. Hajime Inaba.

Behind this shift in viewpoint, there is a recognition of, and emphasis on, the analogy between the two substances (vacuum and ordinary matter) as far as their matter-like (fermion-like) behaviour is concerned, despite their ontological difference: while one is unobservable and virtual and the other is observable and real, they affect each other since they respond in the same way to the exclusion principle.

At the heuristic level, in Nambu's picture the self-energy of a pion depends on whether the latter is propagating through the "vacuum" or through the medium "vacuum + nuclear matter", where both are conceived as *di-mesic* media with different properties. Regarded in analogy with the superposition of two dielectric media, as Nambu suggests, the presence of nuclear matter/*di-mesic* medium "added on" the vacuum modifies the di-mesic properties of the latter.[24] Through this scheme, assuming the nucleus at the roughest as a degenerate Fermi gas, Nambu is able to provide the estimation of the shift in the self-energy of the pion that had originally motivated the model, but also an effective and visual recipe to deal with various nuclear reactions (including pion capture).

More important than these applications, however, is the implication of this view for the nuclear interactions. Notwithstanding the crudeness of the approximation hitherto employed, Nambu's treatment contains an essentially new way to deal with many-body nuclear systems. Rather than trying to extend the potential to higher orders, this scheme works with the simple second-order "korrespondenzmässigen" form of the force potential of Yukawa, while approximately including in it the effects of the non-perturbative many-body interactions via the modification of the self-energy of the pion. In this way, Nambu ingeniously circumvents the disputable way in which quantitative conclusions were drawn in the Yukawa theory, that is, by using the standard concept of potential based on perturbative expansion.

If the core idea of this new treatment is all in there, what Nambu still lacks is the machinery to calculate many-body effects of a nuclear matter that is more complex than just a Fermi gas. To the rescue, supplying the missing elements, will come the collective models on which Tomonaga, Bohm and Pines are independently working at the time of Nambu's speculations and which we laid out in Sects. 2.2.1 and 2.2.2.

3.2.2 Nuclear Matter as a Plasma Medium

In 1951, returning to the questions of nuclear physics from a half-year involvement in cosmic-ray particle physics,[25] Nambu reconsiders the two pathways towards a rigorous relativistic treatment of the nuclear forces—the sought-for "New Formalism"—that he had explored the year before. One had assumed the *deuteron* as its paradigm

---

[24] This picture will be then applied by experimentalists and theorists in Japan to model other nuclear reactions (see, e.g., Machida and Tamura 1951) or to roughly estimate the change of the properties of nucleons due to the nuclear media (see, e.g., Miyazawa 1951).

[25] Here we refer to Nambu's "detour" into cosmic-ray particle physics—a focal point of the experimental and theoretical group in Osaka—after the confirmed discovery, in the mid 1950, of the so-called V-particles (see also Footnote 13).

system and attempted to include in the inter-particle potential high-order effects term-by-term with perturbation theory. The other pathway, stimulated by phenomenological work on the nuclear reactions, had assumed the *nuclear matter* as its paradigm system, considering it as a "di-mesic" medium modifying the "di-mesic" properties of the vacuum, and envisaged to include the high-order effects directly in the propagation properties of the nuclear force carrier (via a modification of its self-energy). The former using the perturbative standpoint—the standard in quantum field theory—and the latter using the simplest possible mean-field standpoint—the standard in solid-state physics—both methods were trying to include the high-order terms which were required to truly characterise nuclear forces. Each with its own obstacle, neither of these two routes to a similar objective had, in fact, managed to completely fulfil it, leading to a halt in both programs. In the former, the problem was the inadequacy of the perturbative approximation. In the latter, assuming the nucleus as a Fermi gas was obviously too drastic an idealisation. Triggered by the collective descriptions of Tomonaga and then of Bohm & Pines, Nambu decides to proceed further on the latter path.

As we have seen, Tomonaga had shown that adding *long-range* interactions to the (non-interacting) particles of a Fermi gas resulted merely in a shift in the energy/frequency of the collective excitations, that is, it amounted to giving "mass" to its previously massless sound waves. That effect suggests to Nambu that perhaps adding short-range interactions to the non-interacting nucleons he had assumed in his Fermi gas model of the nuclear medium could correspondingly result in an additional *mass* of the pion. This clue materialises into a more concrete possibility when "one night in Osaka" in the fall of 1951, Nambu comes across the paper by Bohm and Pines on a new description of metals and plasmas (Bohm and Pines 1951), and quickly realises that this is what his model of nuclear matter would need (Ashrafi 2004; Nambu 2010). As sketched in the previous chapter, via a clever transformation of the coordinates of the single electrons and the electromagnetic field variables of a interacting electron gas, Bohm and Pines had demonstrated that the long-range correlation between electrons induced by the Coulomb interactions leads to longitudinal collective oscillations—an organised sound-wave-like motion of all the electrons in the medium—with a mass. In that same article, the two physicists showed also that these longitudinal oscillations combine with the transverse component of the electromagnetic field to yield massive photons. In this way, as for the longitudinal oscillations, the mass of these photons came to depend on the properties of the gas, in particular its electron density.[26] This concretely implied that, as a result of the collective motion of the electrons and the electromagnetic field that this motion produces, regular photons penetrating the electron gas from the outside become massive photons inside the medium. It also implied that the standard long-range Coulomb forces (originally of the form $1/r$) between the electrons of the medium become screened, short-range forces (of the form $1/r\, e^{-\omega_p r}$) whose *range* depend dynamically on a

---

[26] In the plasma both the longitudinal oscillations and the transverse oscillations (the photons) have an energy-momentum dispersion relation that is affine to that of a massive particle, $E^2 \propto \omega_p^2 + c^2 k^2$, rather than to that of the standard photon $E^2 \propto c^2 k^2$. The mass of the photon in question amounts to the plasma frequency $\omega_p^2 = 4\pi n_0/m$, where n_0 is the electron density of the medium.

property of the gas—see also Sect. 2.2.2. These findings evidenced the qualitative difference between the screening of electromagnetic forces in a dielectric and in a plasma. Unlike in a dielectric, in a plasma to be affected is not the intensity/coupling constant of the electromagnetic force between charged particles, but the range of the force, and thus the mass of the force carrier.[27]

It is between these elements of the Bohm and Pines's picture and his tentative nuclear model that Nambu perceives an analogy which he believes might explain a number of key properties of nuclear interactions: the stability of nuclei and their constant density,[28] and the relative success of single-particle models despite the strength of the interactions. Analogising the nucleus with a plasma—rather than with the (non-interacting) di-mesic medium as Nambu had done earlier—and the (virtual) pions in the nucleus with the photons in the plasma, the pions would acquire an additional dynamical mass as a result of the interaction with the collective degrees of freedom of the nuclear medium. Since that mass would be correlated with the range and depth of the inter-particle nuclear force, this could explain how, inside of a nucleus, such force became weaker and had a shorter range as the nuclear density increased, and vice versa as the density decreased. This would explain why nuclei do not collapse into lumps of infinite density, as well as their observed stability and constant density (see also Kinoshita and Miyazawa 1951). Besides, the lack of strong correlations between the nuclei at short distances that this picture predicted could possibly explain, as it had done for electrons in metals, why nucleons were well described both as independent particles moving in a common potential and as a system of strongly interacting particles moving as a collective. This would justify the success of Bohr's collective model and the otherwise "unreasonable" success of single-particle models (e.g., the shell model).[29] In other words, in addition to the aforementioned properties, the application of the plasma model to the nuclear situation seemed to be able to account for that coexistence between individual and collective phenomena characteristic of the nuclear phenomenology which individual-particle models alone or collective models alone could not account for.

---

[27] As we will see, a decade later through the works of Nambu first and Anderson then, the vacuum of quantum field theory will be recognised as a plasma in this sense.

[28] This is the so-called saturation of the nuclear density and of the binding energy per nucleon. The saturation expresses the experimental fact that both these quantities are practically independent from the number of nucleons, rather than increasing proportionally with the number of nucleons as they should have according to the Yukawa model (and even just according to a simple Fermi gas model of the nucleus).

[29] In general, Bohr's compound model, the prototype of collective models of the nucleus proposed in 1936 (see Sect. 2.2.1), was successful with most of the experimental observations (Amaldi 2004). However, it could not explain, for instance, why some nuclei with "magic" numbers (2, 8, 20, 28, etc.) were particularly stable, much like the atomic configurations of noble gases, a fact which instead had been explained in 1950 by the shell model of Goeppert-Mayer and Hans Jensen, where the nucleons are assumed to move under a central force (Goeppert-Mayer 1950).

The Collective Descriptions in the Language of Quantum Field Theory

The outcome of Tomonaga's and Bohm and Pines's collective descriptions encourage Nambu's earlier attempt at the nuclear many-body problem, and will lead him and his collaborator Toichiro Kinoshita to a coherent and general treatment for many-body systems, as we will see. Such a treatment incorporates the assumptions and results of the collective descriptions, but at the same time it aims at reformulating them by making more integral use of the language of quantum electrodynamics. "The essential idea of my treatment is the same as yours" writes Nambu to David Pines with his characteristic understatement (30th September 1952), "but I make use of the recently developed ideas of quantum electrodynamics: the interaction representation and the renormalisation" (Nambu 1952a). As we have described in Sect. 2.2.2, Bohm and Pines, too, had drawn ideas and techniques from renormalised quantum electrodynamics, but in the formulation which Schwinger and Tomonaga had given it. Nambu instead aims at employing as a framework the other formulation of quantum electrodynamics, that of Feynman and Dyson (1949). In common with Tomonaga and Bohm & Pines, Nambu assumes that the bulk of the interaction energy in a many-body system can be expressed through a change of the properties of the force carrier—in the case of plasma, the mass of the photon field. In contrast to them, however, this change in particles' properties is not to be achieved by going over to the collective description, but more naturally using the particle perspective adopted in Feynman and Dyson's formulation, its physical picture and tools (the Green's functions) (Nambu 1952b). In what is, in fact, a first attempt to arrive at a fully-fledged many-body quantum field theory—which could be then applied to nuclei and plasmas, among others—Nambu's idea is to express the change in particles' properties as the outcome of a "second-order" renormalisation process by using the structural analogy that he had redrawn between the vacuum and the Fermi sea. Nambu practically does that by taking more literally the analogy between the renormalisation procedure and the Hartree-Fock method which had previously inspired Tomonaga (see Sect. 2.1.3).

Dealing with the many-electron problems in its various applications, the basic idea behind the Hartree-Fock method, proposed twenty years earlier in the context of atomic physics, was to first formally replace the (*non-linear*) electron-electron interaction energy h_{int} with a generic, *linear* and single-particle, effective pseudo-potential $\phi(\Psi)$. The latter essentially expressed the common potential to which each electron in an atom is subject to due to the configuration of surrounding electrons $\Psi(\phi)$, but assuming this to be the configuration of the "free" electrons (that is, the product of their wavefunctions considered without the electron-electron interactions). The values of these two co-dependent variables, the effective potential and the total electronic configuration, were then determined by means of a specific self-consistent equation so as to best reproduce with the single-particle potential the original interaction h_{int}. As Hartree had proven, the values so obtained were in fact the optimal ones in two senses. Firstly, the final self-consistent potential ϕ completely absorbs the non-perturbative part from h_{int}, leaving only a perturbative residual electron-electron interaction energy h'_{int}. Secondly, the corresponding electron wave-functions are the free single-particle wavefunctions closest to the many-body case.[30]

Cleverly seeking for a field theoretical calque of this procedure, Nambu considers the inter-particle scattering processes contained in the interaction energy of quantum electrodynamics H_{int}. These are the electron-electron (Møller scattering) and electron-photon (Compton scattering) processes, respectively:

$$\bar{\psi}(x)\psi(x)\bar{\psi}(x')\psi(x')$$
$$\bar{\psi}(x)\psi(x')A(x)A(x')$$

Nambu aims at giving an effective description of them from the single-particle perspective, that is, to transform these non-linear terms into simpler linear terms proportional to $\bar{\psi}(x)\psi(x')$ and $A(x)A(x')$, and that are therefore homogeneous to the electron and photon free propagators, respectively. Nambu's idea is to do that by taking the expectation values of the scattering terms with respect to the configuration of electrons in the ground state $\langle...\rangle_\Psi$, in this way obtaining:

$$\langle\bar{\psi}(x)\psi(x')\rangle_\Psi A(x)A(x') := \chi(x,x';\psi)A(x)A(x')$$
$$\{\langle A(x)A(x'')\rangle_\Psi \langle\bar{\psi}(x'')\psi(x'')\rangle_\Psi + \langle A(x)A(x')\rangle_\Psi\}\bar{\psi}(x)\psi(x') := \phi(x,x';\psi,A)\bar{\psi}(x)\psi(x').$$

In this way, the potentials (or mean fields) χ and ϕ that appear in the equations formally incorporate on average all the scattering processes experienced by, respectively, a photon due to the other electrons, and an electron due to the other electrons as well as photons in the plasma. Given that the quantum fields of electrons and photons determine the two potentials, and these potentials in turn determine the quantum fields, the optimal values for both the potentials and the quantum fields are found by solving two self-consistent coupled equations.

By the way they are calculated and by the function they absolve, the potentials χ and ϕ are essentially Hartree potentials for the photons and the electrons, respectively. But since the scattering terms, so linearised, respectively have the form of photon and electron self-energies, in which the Hartree potentials χ and ϕ play the role

---

[30] In the Hartree-Fock method, the self-consistent equation relating the potential $\phi(\Psi)$ with the final electronic configuration Ψ was derived by means of a variational principle. More specifically, this was done by imposing on the variation the condition:

$$\delta \langle h'_{int}\rangle_\Psi := \langle(h_{int} - \int\int\phi\,\bar{\psi}(x)\psi(x')dx\,dx')\rangle_\Psi = 0.$$

As Nambu notices, this implies the condition $\langle h'_{int}\rangle_\Psi = 0$, which actually amounts to requiring that the electron "self-energy"—in this case, the energy due to the scattering with the surrounding real electrons—arising from the interaction h'_{int} shall vanish. In this sense, the Hartree-Fock procedure can be seen as the non-perturbative analogue of renormalisation. This argument will be explicitly laid out in Kinoshita and Nambu (1954), but already underpins the analogy between renormalisation procedure and Hartree-Fock method drawn by Tomonaga (see Sect. 2.1.3).

of mass-like coefficients, Nambu's procedure allows one in effect to calculate the *additional* self-energies that electrons and photons acquire in the plasma medium. Introducing these additional finite "self-consistent self-energies" in the many-body Hamiltonian, and employing the renormalisation procedure established in quantum electrodynamics, one manages to subtract from H_{int} the non-perturbative part of the many-body interaction and include it in the starting approximation by properly redefining new free electrons and photons (more technically said, by going over to a new interaction representation). Like in the Hartree-Fock, this leaves a residual interaction energy H'_{int} that can be treated perturbatively.

In this way, the renormalised electrons and photons of quantum electrodynamics are the bare particles which get renormalised a second time by their interaction with the medium to become, as they are called, quasiparticles, with different, modified propagation properties.[31] As Nambu writes to Pines, it is in this sense that one can interpret the ground state of the solid—the state in which all the "negative" quasiparticle levels (below the Fermi energy) are occupied—as an "apparent vacuum" in analogy with the vacuum of standard quantum electrodynamics, and thus also interpret quasiparticles as renormalised particles. As from the vacuum of Dirac, pairs of quasiparticles and quasiholes (respectively above and below the Fermi energy) can be excited from this "apparent vacuum".

Reached through the elaboration of Dirac's view of the vacuum as a medium, Nambu's view of the ground state of a solid medium as an "apparent vacuum" and of quasiparticles as its excited states will be the crucial premise of his reformulation of superconductivity first and then of the discovery of spontaneous symmetry breaking.

The Chimera of the Nuclear Problem

Nambu's blueprint for the reformulation of Bohm & Pines' collective description of plasmas here laid out appears, in its essential elements, already in the letter that Nambu addresses to Pines upon his arrival in Princeton on September 30, 1952 (Nambu 1952a). Nambu's objective was though not just to model plasmas, but more generally to offer a "collective description of many-particle systems" or, as he would alternatively call it, "a generalised theory of Hartree fields". This would be inspired

---

[31] It is worth remarking that if this procedure is indeed a non-perturbative analogue of the standard renormalisation in quantum electrodynamics, it presents three significant disanalogical elements. (i) The interactions of "bare" quasiparticles with the real medium result in self-consistent self-energies which are finite and calculable. Unlike in quantum electrodynamics, these are not infinite quantities compensating for bare infinite parameters to produce the properties observed experimentally and inputted in the theory. (ii) Since this is not based on a perturbative procedure, a non-zero parameter can result also from a vanishing bare parameter—e.g., a photon with non-zero mass from a massless bare photon (see also reflections in Sect. 5.1.1). (iii) The particles are modified in their properties not only by the "few-body" local interactions with the vacuum, but by the local and non-local interactions with all the other many bodies in the solid.

by the collective descriptions of plasma, but would aim at modelling various other phenomena, including the properties of nuclei. As Nambu writes at the beginning of his letter:

> I am interested in [your papers on the "collective description" of many particle systems] because Dr. Hayakawa [...] and I are now working together on [...] the theory of plasma oscillations and its generalisation. Indeed I was suggested by your papers (and those by Bohm and Gross) to investigate plasma oscillations as a preliminary to many important problems, such as the properties of nuclei, superconductivity and superfluidity, origin of the solar noise, Heaviside layers, origin of cosmic rays, and the theory of discharge.

The variety of listed phenomena notwithstanding, as the working notes and paper drafts of the period betray, Nambu's main target is in fact what in the notes he refers to as the "collective model of nucleons and mesons" (Nambu, 1953, 1954). As we know, a satisfactory non-perturbative formulation of the Yukawa theory and the *ab initio* explanation of the nuclear properties were the goals that had originally motivated Nambu's endeavour. At the time of the letter, Nambu still hopes that his collective description could do that and, fuelled by the prospect, he devotes the whole period at the Institute for Advanced Studies in Princeton—between October 1952 and the June 1954, in collaboration with his fellow Toichiro Kinoshita—to the systematic formulation of his collective description, and its application to the nuclear problem. The general formulation would be submitted to *Physical Review* at the end of 1953 (Kinoshita and Nambu 1954). After mentioning the "interesting examples which may advantageously be handled by the present formulation", as far as the nuclear problem the article though concedes that:

> In the realm of nuclear physics, the method seems less promising because of the strong interaction and short-range character of the nuclear forces or meson field which makes nuclei look like liquid in some respects. Nevertheless, it would turn out helpful in investigating some important features of the nuclear forces, such as the many-body forces, saturation, and the shell structure, which have not yet been fully understood. These problems will be studied in forthcoming papers.

Despite the efforts, the projected follow-up article "On the saturation of nuclear forces" along with the sought-for application to the nuclear problem would never work out (Nambu, 1953, 1954). As anticipated by the two physicists, the main impediment indeed turns out to be the strength and short range of the nuclear forces as compared to the electromagnetic forces acting in plasmas. The long and detailed Physical Review communication by Nambu and Kinoshita will remain virtually unknown and barely cited in the years to come, but constitutes the premise of the field theoretical treatment of superconductivity by means of which spontaneous symmetry breaking will be discovered. This is how the nuclear problem once again escaped the complete and satisfying formalisation that Nambu had in mind to remain the chimera it had hitherto been. Nambu's efforts and obsession for it were not in vain though, as they had led him somewhere from which an unexpected part of the journey would begin.

References

Amaldi, U. (2004). Nuclear physics from the nineteen thirties to the present day. In *Enrico Fermi: The life and legacy*, pp. 151–176. Springer.

Ashrafi, B. (2004). Interview of Yoichiro Nambu by Babak Ashrafi on 2004 July 16. www.aip.org/history-programs/niels-bohr-library/oral-histories/30538.

Bohm, D., & Pines, D. (1951). A collective description of electron interactions. i. Magnetic interactions. *Physical Review 82*(5), 625.

Brown, L. M., & Nambu, Y. (1998). Physicists in wartime Japan. *Scientific American, 279*(6), 96–103.

Dirac, P. A. M. (1930). A theory of electrons and protons. *Proceedings of the Royal Society of London. Series A, Containing Papers of a Mathematical and Physical Character 126*(801), 360–365.

Dirac, P. A. M. (1931). Quantised singularities in the electromagnetic field. *Proceedings of the Royal Society of London. Series A, Containing Papers of a Mathematical and Physical Character 133*(821), 60–72.

Dyson, F. J. (1949). The radiation theories of Tomonaga, Schwinger, and Feynman. *Physical Review, 75,* 486–502.

Furlan, S., & Gaudenzi, R. (2021). Far from the particle crowd: Shugyosha Nambu and Michizane Wheeler. *Proceedings of 40th Congress of the Italian Society for the History of Physics and Astronomy,* 147–153.

Goeppert-Mayer, M. (1950). Nuclear configurations in the spin-orbit coupling model. i. Empirical evidence. *Physical Review, 78,* 16–21.

Hayakawa, S. (1991). Sin-Itiro Tomonaga and his contributions to quantum electrodynamics and high energy physics. *Progress of Theoretical Physics Supplement, 105,* 157–167.

Hoddeson, L., Schubert, H., Heims, S., & Baym, G. (1992). Collective phenomena. *Out of the crystal maze: Chapters from the history of solid state physics.* Oxford University Press.

Kinoshita, T. (1991). Personal recollections, 1944–1952. *Progress of Theoretical Physics Supplement, 105,* 116–119.

Kinoshita, T., & Miyazawa, H. (1951). On the saturation of nuclear forces (in Japanese). *Soryushiron Kenkyu, 3*(5), 154–157.

Kinoshita, T., & Nambu, Y. (1954). The collective description of many-particle systems (a generalized theory of Hartree fields). *Physical Review, 94*(3), 598.

Kono, H. (2020). Ryogo Kubo in his formative years as a physicist. *The European Physical Journal H, 45*(2), 175–204.

Lax, M., & Feshbach, H. (1951). Production of mesons by photons on nuclei. *Physical Review, 81*(2), 189.

Low, M. (2005). *Science on the International Stage: Hayakawa,* pp. 169–196. Palgrave Macmillan US, New York.

Low, M., Nakayama, S., Yoshioka, H., et al. (1999). *Science, technology and society in contemporary Japan.* Cambridge: Cambridge University Press.

Machida, S., & Tamura, T. (1951). On the photo-meson production from heavy nuclei. *Progress of Theoretical Physics, 6*(3), 437–439.

Marshak, R. (1952). *Meson Physics.* International series in pure and applied physics. McGraw-Hill.

Miyazawa, H. (1951). Deviations of nuclear magnetic moments from the Schmidt lines. *Progress of Theoretical Physics, 6*(5), 801–814.

Nambu, Y. (1949a). The level shift and the anomalous magnetic moment of the electron. *Progress of Theoretical Physics, 4*(1), 82–94.

Nambu, Y. (1949b). On the method of the third quantization. *Progress of Theoretical Physics, 4*(3), 331–346.

Nambu, Y. (1950). A note on the eigenvalue problem in crystal statistics. *Progress of Theoretical Physics, 5*(1), 1–13.

Nambu, Y. (1950). Force potentials in quantum field theory. *Progress of Theoretical Physics, 5*(4), 614–633.

Nambu, Y. (1950c). Nambu, Yoichiro. Papers, [Box 4; Folder 2], Special Collections Research Center, University of Chicago Library, Chicago, USA.

Nambu, Y. (1950). Pi-meson self-energy (in Japanese). *Soryushiron Kenkyu, 2*(2), 170–174.

Nambu, Y. (1952a). Correspondence from Yoichiro Nambu to David Pines, 30 September 1952 (unprocessed), David Pines Papers, University of Illinois at Urbana-Champaign, Urbana, USA (Courtesy of the University of Illinois Archives).

Nambu, Y. (1952b). Nambu, Yoichiro. Papers, [Box 4; Folders 6], Special Collections Research Center, University of Chicago Library, Chicago, USA.

Nambu, Y. (1953, 1954). Nambu, Yoichiro. Papers, [Box 12; Folder 3, 6], [Box 24; Folders 4-11], Special Collections Research Center, University of Chicago Library, Chicago, USA.

Nambu, Y. (1991). Summary of personal recollections of the Tokyo group. *Progress of Theoretical Physics Supplement, 105,* 111–115.

Nambu, Y. (1995). Research in elementary particle theory. In *Broken Symmetry: Selected Papers of Y. Nambu*, pp. vii–xiv. World Scientific.

Nambu, Y. (2010). Energy gap, mass gap, and spontaneous symmetry breaking. In *BCS: 50 years*, pp. 525–533. World Scientific.

Nambu, Y. (2016). Reminescences of the youthful years of particle physics. In *Memorial volume for Y. Nambu*, pp. 143–159. World Scientific.

Nambu, Y., Nishijima, K., & Yamaguchi, Y. (1951). On the nature of V-particles, I. *Progress of Theoretical Physics, 6*(4), 615–619.

Nishijima, K. (1951). On the adiabatic nuclear potential. I. *Progress of Theoretical Physics, 6*(5), 815–828.

Oneda, S. (1991). Personal recollections on weak interactions during the 1950's and early 1960's. *Progress of Theoretical Physics Supplement, 105,* 123–128.

Onsager, L. (1944). Crystal statistics. I. A two-dimensional model with an order-disorder transition. *Physical Review 65*(3–4), 117.

Orear, J., & Fermi, E. (1950). *Nuclear physics: A course given by Enrico Fermi at the University of Chicago*. University of Chicago Press.

Tanikawa, Y. (1948). Recent developments in meson theory (in Japanese). *Kagaku, 18,* 439.

Tomonaga, S. (1952). Correspondence from Sin-Itiro Tomonaga to Robert Oppenheimer, 22 January 1952, Director's Office: Member Files: Box 101: Nambu, Yoichiro, Shelby White and Leon Levy Archives Center, Institute for Advanced Study, Princeton, NJ, USA.

Weart, S. (1992). The solid community. In *Out of the crystal maze: chapters from the history of solid state physics*. Oxford University Press.

Weisskopf, V. F. (1949). Recent developments in the theory of the electron. *Reviews of Modern Physics, 21*(2), 305.

Wentzel, G. (1947). Recent research in meson theory. *Reviews of Modern Physics, 19*(1), 1.

Wright, A. S. (2016). A beautiful sea: PAM Dirac's epistemology and ontology of the vacuum. *Annals of Science, 73*(3), 225–256.

Yamaguchi, Y. (1951). The phenomenological analyses of mesonic processes. *Progress of Theoretical Physics, 6*(5), 772–787.

Yukawa, H. (1935). On the interaction of elementary particles. i. *Proceedings of the Physico-Mathematical Society of Japan. 3rd Series 17,* 48–57.

Chapter 4
Elementary Particles as Excitations of a Solid Medium. Is the Universe a Superconductor?

In this chapter we trace the steps leading up to a theoretical understanding of super-conductivity through the formulations given first by Bardeen, Cooper and Schrieffer, and then by Bogolyubov. We then see how Nambu resolves the problematic violation of charge conservation/gauge invariance left unresolved by both formulations; and how, in doing so, he uncovers in superconductors what would be called spontaneous symmetry breaking. Via an analogy between his treatment of superconductors and elementary particle theory, Nambu recognises the same mechanism as being at work in the universe at large and as being responsible for the mass of elementary particles. This picture introduces in physics the concept that the particles and bosons that we observe are not excitations of an inert, fully-symmetric vacuum, but excitations of a vacuum endowed with superconducting-like properties, different and yet logically connected manifestations of an underlying interaction. This shows how the analogy between the world of low-energy phenomena and that of high-energy phenomena has made intelligible a general principle underlying both that could hardly be disclosed in either of the two alone.

The physics of the post-war period spanning the decade '50–'60 is an extremely rich and fruitful period of 20th century science. Among others, it is traversed by two important simultaneous discoveries: the *theoretical* discovery of the microscopic mechanism of superconductivity in 1957, and the *experimental* discovery—right in the same period—of the first violation of a symmetry that was believed to be foundational and inherent to the physical world, the mirror symmetry (also called parity symmetry). Albeit belonging to two entirely different orders of phenomena, both were somehow about symmetry and symmetry breaking. A more subtle and puzzling matter emerging from its theoretical treatment, superconductivity seemed to violate gauge symmetry, an *internal* symmetry which is related to the conservation of electric charge, and which was thought to be an equally inviolable principle of the physical world and one that had been the base of the previous hundred years of theoretical physics. While the principle indeed turned out to be upheld, that occurred in a way that had never before been recognised. On a parallel front of physics, the violation of mirror symmetry discovered in particle physics had stimulated the search for a different, "higher" symmetry. Such a symmetry, identified in chiral symmetry, however, seemed to be incompatible with the very presence of massive particles—incompatible unless, as Nambu correctly intuited in a conceptual leap, something analogous to the superconductor was at work in the universe at large, and analogous problems could be treated with the same means. This was how the unexpected bridging of these two parallel orders of problems and worlds of phenomena enabled the recognition of the underlying principle of spontaneous symmetry breaking. This is the heuristic path that we will reconstruct in the following, beginning with the way superconductivity was discovered.

4.1 Constructing Superconductivity

At the appearance of a phenomenon christened "supra-conductivity"—before being called superconductivity—Heike Kamerlingh Onnes and his laboratory assistants seem to have been caught rather unprepared. Tracking the electrical resistance of a wire of pure mercury as it cooled down to lower and lower temperatures, they realised that the value of the resistance fell below the sensitivity of the measuring instrument. In fact, as Kamerlingh Onnes would write, the resistance "becomes zero, or at least differs inappreciably from that value" (Kamerlingh Onnes 1911).

The young Kamerlingh Onnes had begun his experimental work at the University of Leiden in 1882 with a clear poetics in mind[1]:

> In my opinion, in the experimental practice of physics, the pursuit of quantitative research, that is to say, detecting the numerical relationships of measure in phenomena, must be at the forefront. *From measuring to knowing*, I would like to write as a motto at the entrance of every physical laboratory.

The practical implementation of the motto *from measuring to knowing*[2] necessarily meant being able to manufacture instruments, possibly in house and of very good quality. Upon realising this key factor of the experimental activity of the time, Onnes had also realised that such manufacturing in fact took the largest part of the experimenters' time, and consequently that the time available for research and speculation would be stretched if the physicists were to closely collaborate with trained technicians in this work (something which would become a common practice in experimental physics till today). With this rationale and a systematic investment in

---

[1] This is an excerpt of the inaugural lecture "De beteekenis van het quantitatief onderzoek in de natuurkunde" (The significance of the quantitative research in the natural science) delivered (in Dutch) by Kamerlingh Onnes in November 1882 in Leiden, and reported in the biography written by Dirk van Delft (2005). It is perhaps worth remarking that the University of Leiden was charged at that time with the longest scientific and philosophical tradition of the Netherlands; and was home of other young experimentalists and theorists like Peter Zeeman and Hendrik Lorentz.

[2] The original Dutch for it is "Door meten tot weten" which has a more pleasant assonance.

instrument making as well as skilled technicians, over the years Kamerlingh Onnes had set up a large, organised, and efficient laboratory whose focus was the investigation of the physics of low temperatures (Eckert et al. 1992; Reif-Acherman 2004).[3]

Two were in particular the phenomena which Kamerlingh Onnes was determined to investigate: the behaviour of the simple real gases—and in particular the testing of the van der Waals law—and that of the electrical conductivity of metals in the more interesting and yet uncharted low temperature realm (Hoddeson et al. 1992). Access to both these behaviours depended on the practical possibility of liquefying these gases. For the former investigation, the object of study would be the gas itself and its transition to liquid state; while for the latter the liquefied gas would function as a means. Used as the refrigeration fluid in contact with the metal sample, the liquefied gas would act as a heat sink at a constant temperature through which the sample could be cooled down. That temperature, being the temperature of the liquid-to-gas phase transition, depended on the pressure that the liquid was subjected to as well as on the properties of the gas used. This meant that, in order to explore the electrical conductivity at the lowest possible temperatures, one had to liquefy the gas with the lowest critical temperature. Owing to the smallest intermolecular forces among the known gases, the "noblest" helium gas, with its extraordinarily low boiling point at standard pressure of 4.2K, was at once the best candidate and hardest challenge. Air first, and then separately oxygen, nitrogen, and hydrogen (with the respective boiling points of about 90K, 77K, 21K) had been liquefied at different moments in time from the 1870s onward independently by Louis-Paul Cailletet (oxygen), Carl von Linde (air and nitrogen), and James Dewar (hydrogen).

Having lost the battle against Dewar for the first drops of liquid hydrogen ever obtained, Kamerlingh Onnes instead embarked on the design and construction of an efficient and powerful hydrogen liquefier that could systematically produce liquid hydrogen in much larger quantities than Dewar had done (Dahl 1984; van Delft 2005). This entailed an eight-year-long fine work of engineering, but constituted the key to the process that culminated, on July 10 1908, in the liquefaction of helium. The hydrogen liquefier allowed first for studying the properties of helium and understanding whether there was a chance of liquefying it, served as a master model for the helium liquefier, and supplied the liquid hydrogen necessary for the pre-cooling of helium in the helium liquefier. As Kamerlingh Onnes and his crew discovered, the liquefaction occurred at a temperature of about $-269°$ C (corresponding to 4.2 K). Starting from that temperature and reducing the pressure above the liquid, a temperature of about 1.5K was attained (van Delft 2008). At that time, that was certainly the coldest point on Earth and, as far as human mind can judge, in the universe at large.

---

[3] Eckert et al. (1992) and van Delft (2005) interestingly emphasise how, in terms of material and intellectual resources employed as well as the strategies, these efforts can be regarded as the "big science" of the time.

The positive feedback loop of scientific knowledge and instrument engineering that enabled Kamerlingh Onnes and his crew to obtain liquid helium unlocked a door to the physics of metals in a completely uncharted temperature range. The theories and laws proposed by Eduard Riecke and Paul Drude, and later Hendrik Lorentz, on the electrical and thermal properties of metal and metal conduction were successful, but how these depended on temperature and whether they would hold at low temperatures was still much of an hypothesis (Gavroglu 1995). Now, verifying, rejecting or correcting the speculations was in fact the spirit behind the measurements on the resistance of metals that Kamerlingh Onnes hurriedly began as soon as he obtained liquid helium. In what was a continuation of the work initiated with his *promovendus* Jakob Clay into the newly obtained first-ever-explored temperature regime (from 20 K to about 1 K), Kamerlingh Onnes probed the resistivity of various metals of different purity while cooling them down across the temperature range. In these studies performed over the course of 1911, he found, for instance, that the resistivity in general was *continuously* decreasing until it approached a constant— rather than decreasing to a minimum and then increasing again, as Lord Kelvin had predicted—whose value was lower for metals with fewer impurities and practically zero for pure metals (Hoddeson et al. 1992). Kamerlingh Onnes did not, however, encounter anything vaguely resembling the known or imagined when he more accurately measured mercury in the neighbourhood of 4K: the resistance of pure mercury, as he noted in his log book, was "disappearing" and reappearing *abruptly* in an interval of hundredths of a degree kelvin around a "threshold" value as the sample was first cooled and then reheated across the threshold (Hoddeson et al. 1992).

Completely different from a trend that, whatever form it might have taken, was assumed to be continuous in any proposed and conceivable theoretical model, this "state" characterised by the loss of electrical resistance was the following year found by Kamerlingh Onnes also in samples of lead and tin. Experimenting on the reaction of these superconductors to other perturbations, Kamerlingh Onnes discovered that, at least in some metals, a rather weak magnetic field was enough to cause the disappearance of superconductivity. Similarly, a current passing through the superconducting wire and exceeding a certain (rather high) threshold value—as had occurred for temperature—caused the sudden rise of wire resistance and the breaking of the superconductive condition. In the cautious attempt to "first [...] refer the [new] phenomena as much as possible to the known ones" as long as "the contrary is not experimentally proved" Kamerlingh Onnes ascribed the breaking of the superconducting state to ohmic heating effects (Kamerlingh Onnes 1913).[4] To back this assumption, he set out to perform experiments introducing gold and cadmium impurities in the pure mercury samples (Hoddeson et al. 1992). Very surprisingly, in stark contrast to what happened with the same metal in the normal state and against his original belief, the amount and type of impurities did not seem to affect the characteristics of the superconducting state, suggesting thereby the total absence of current related heating effects and the essential *invalidity* of Ohm's law in this newly discovered "metal-

---

[4] Ohmic heating is the heating due to the scattering of electrons with impurities that occurs in any standard conductor obeying Ohm's law.

lic" state.[5] This remarkable absence of a "micro-residual resistance", as Kamerlingh Onnes had called the object of his search, was firmly established by a further experiment he performed in 1914 (Kamerlingh Onnes 1914). This consisted in joining the ends of a superconducting lead coil in which a current was passing, and observing it running through the coil without externally applied voltage—as though, he writes, it were a "molecular current as imagined by Ampere"—while tracking its extremely slow decay in time. The molecular current that Kamerlingh Onnes here suggestively analogises to the persistent current in a superconducting ring was the "current" generated by electrons in their *stationary* frictionless orbits around the atomic nucleus, which Ampere had thought of as being responsible for magnetism. In fact, in a real superconductor, differently than in the atomic case, non-idealities led to a small power dissipation and a corresponding decay of the current in time. Measuring this dissipation, Kamerlingh Onnes' experiment allowed him to fix the upper limit for the residual resistance of the superconducting state to a value of 10^{-10} times the resistivity of the same material in the normal state (at the reference value of $0°$ C) (Kamerlingh Onnes 1914). It was as if, with superconductivity, a piece of the ideal world, a *perpetuum mobile*, had shown up in the real world!

The astonishingly low resistivity, the peculiar mixture of fragility against magnetic fields and robustness against currents and impurities exhibited by superconductors defied any simple explanation. The main reason for this radical unexplainability, not clear in the immediacy of the discovery (Hoddeson et al. 1992), lay in the fact that a normal metal upon becoming superconducting enters into a state which has little, or nothing, to do with the metallic state it comes from. This was a new state to describe which were required conceptual tools that had not yet been invented. These low-temperature pioneers were unknowingly dealing with a macroscopic quantum state emerging as a consequence of a rather exotic phase transition, years before either quantum mechanical behaviour or phase transitions would be theorised.

The understanding of this peculiar macroscopic and collective quantum state slowly took off two decades after Kamerlingh Onnes' discovery, and was determined by a series of conceptual shifts, the last of which is Nambu's discovery of spontaneous symmetry breaking. The first of such shifts was a shift away from the somewhat dogmatic and partly unconscious assumption that the superconductor was a metal with infinite conductivity, an assumption that had been impressed by the authoritative experiments of Kamerlingh Onnes and his collaborators in Leiden, and would continue to linger in the years to come.

---

[5] To complicate the puzzle, as would become clearer in further experiments, these impurities were, if anything, *strengthening* the superconducting state against magnetic fields.

4.1.1 Superconductivity Is Most of All Perfect Diamagnetism, or the Consequences of a Perspective Shift

As we have mentioned, one of Kamerlingh Onnes' research objectives was to study the electrical conduction properties of metals at low temperatures in order to experimentally test the existing theoretical models. Given this focus on electrical properties, it was therefore quite natural for him to identify the state of "supra-conductivity" that some metals entered into at sufficiently low temperature as uniquely defined by a shift in resistivity, which left the magnetic properties essentially unchanged (Gavroglu and Goudaroulis 1984). Framing the superconductor as an "ideal" metal—that is, using the zero resistivity as the defining property, and thus considering Ohm's law as valid—had other theoretical implications for the response to electromagnetic fields. If, in particular, the superconductor became superconducting in the presence of an external magnetic field, it should have trapped the field inside of it and kept it even after turning the external field to zero.[6] That implied that the same superconductor subject to the same conditions, but with a different history, would have a different state. The fact that the state of a superconductor depended upon its past history had, in turn, a negative implication for its thermodynamic description. It meant that the superconducting state was not a state of thermodynamic equilibrium and, as a consequence, one could not treat the transition to superconductivity as one treated all other known phase transitions. To be sure, according to the set of experiments performed by Kamerlingh Onnes and Tuyn in 1922, this trapping/"freezing" of the magnetic field seemed to occur in some cases (Gavroglu and Goudaroulis 1984; Tuyn and Kamerlingh Onnes 1991). But the experiments themselves were very challenging to perform—they were, among other things, quite sensitive to the shape of the samples (Hoddeson et al. 1992). As other experimenters would later reveal, in that case the investigation and interpretation of the results given by the two scientists had not been as thorough as in other cases.

The phenomenology that had emerged from the numerous experiments performed in the Leiden laboratory from the discovery of superconductivity on was given a mathematical formulation by Hendrik Lorenz, and then in a work by Becker, Heller, and Sauter in the early thirties (Becker et al. 1933). The four Dutch physicists assumed the superconductor to be the "ideal" metal that it appeared to be and studied the consequences of Ohm's law in the limit of infinite conductivity. In this limit, the electrons would constantly accelerate under the applied electric field E, rather than having, on average, a constant velocity as happens in a standard metal. Mathematically, this was expressed by imposing that the time derivative $\partial_t j$, rather than the current density j, be proportional to the applied electric field E. Inserting such a modified Ohm's law

---

[6] This descended from a rather straightforward application of Maxwell's laws to a perfect conductor: the change of magnetic flux through a material generates an electromotive force which, in turn, causes electric currents to flow that resist that change (Faraday-Lenz law). If the material has zero electrical resistance, such currents are always so as to perfectly counterbalance the change in external magnetic field. This would result in a fixed magnetic flux through the superconductor which cannot be changed, i.e., which remains "trapped" or "frozen in".

in Maxwell's law of induction—relating the electric field with the magnetic field—they obtained an inhomogeneous differential equation describing how an external magnetic field H_{ext} would behave in the interior of the superconductor (Becker et al. 1933). The equation predicted a rapid exponential decay of the magnetic field down to a non-zero value H_0 inside the superconductor.[7] This field represented the field present in the metal when it was last cooled below the superconducting transition temperature. The mathematical-phenomenological framework of Becker, Heller, and Sauter therefore appeared to be consistent with the "frozen" magnetic field that had been observed, and theoretically corroborated that, if superconductors were defined as perfect conductors, they could not be treated thermodynamically.

That fact and the impediment that resulted were being challenged by the systematic experiments that Walter Meissner and Robert Ochsenfeld were conducting in Berlin at the same time. The result of these experiments, released in the wicked year 1933 in a one-page communication entitled "A new effect upon entering superconductivity" (Meissner and Ochsenfeld 1933), started gnawing away the deceptive link that had been assumed between superconductivity and the metallic state.[8] The mechanical engineer and physicist Walter Meissner had been appointed director of Berlin's Physikalisch-Technische Bundesanstalt in 1922. There, within a few years, he had set up a large facility for helium liquefaction and started to do research at low temperatures. In collaboration with Ochsenfeld, Meissner focused on experimentally characterising the electromagnetic response of superconductors. Under the suggestion of Max von Laue, the two scientists planned to probe whether the super-current in a superconductor flows on the surface or in its interior (Gavroglu 1995; Goodstein and Goodstein 2000). They intended to do that by measuring the intensity of the externally-applied magnetic field in the space between two cylinders (pure tin the first pair and lead the second pair) as they were being cooled, in a helium bath, below the critical temperature. The result of the experiment is well-known nowadays. Upon crossing the superconducting transition temperature, the two scientists somewhat cautiously concluded, "in contrast to [previous views]... the force line distribution in the outer environment of the superconductors changes and becomes almost as it would be expected if the superconductor had a zero [magnetic] permeability" (Meiss-

---

[7] The field was shown to decay exponentially as a function of the distance from the surface—where it was fitted to the value H_{ext}—to reach the constant value H_0 within the very short distance given by the so-called penetration depth $\lambda = c\sqrt{\Lambda/4\pi}$, where $\Lambda = m_e/ne^2$ is a property of the given material (m_e stands for the electron mass, e for its charge, and n for the electron density).

[8] Playing a bit with counter-factuality, we remark that the link most likely would not have been made if superconductivity had been first discovered in insulators. In this sense, that link was the result of a contingency, in fact a rather complex one and thereby a posteriori justified. The link certainly depended on the theoretical horizon at that particular moment of history and was reinforced by the erroneous interpretation of (delicate) experimental results, but it also depended on the physical measuring capabilities. Hardly anyone would have thought of measuring the resistance of an insulator down to a few degrees Kelvin. The obvious reason was that the resistivity was assumed to increase; and second because it was hard to measure it without passing a current. We might say that analogous beliefs about the positive correlations between the purity of a material and its conductivity would later "slow down" the discovery of disordered superconductors.

ner and Ochsenfeld 1933).[9] In other words, the results quite convincingly showed that the two cylinders completely expelled the magnetic field from their interior upon becoming superconducting.

For those few who immediately accepted the experimental result, seeing in the "new effect" discovered by Meissner and Ochsenfeld a confirmation of their beliefs and intuition, the experiment had two important and related implications for the thermodynamics and the electrodynamics of superconductors (Sardella 2012). With regard to the former, the result convincingly suggested that a superconductor, in contrast to the then widely held belief, admitted only one unique thermodynamic state of equilibrium. This state was characterised by the complete absence of a magnetic field in the interior of the superconductor rather than by the presence of any trapped magnetic field, a fact which quickly re-established the possibility of giving superconductivity a thermodynamic treatment. This brought Paul and Tatiana Ehrenfest, and independently Lev Landau, to hypothesise that superconductivity was indeed the result of a phase transition as was, for instance, the ferromagnetic state. It also led Cornelius Gorter and Hendrik Casimir to derive on the basis of purely thermodynamic considerations the thermal properties of superconductors and the critical magnetic field (Sardella 2012). On the side of electrodynamics, Meissner and Ochsenfeld result invited the shift from the older and prejudicial definition of a superconductor as a perfect conductor to its definition first and foremost as a perfect *diamagnet* (Gavroglu 1995).

The physicists Fritz and Heinz London were the first who perspicaciously worked out and presented in 1934—a few months after the result—the consequence of the suggested perspective shift for the electrodynamics of superconductors. The two brothers, German political refugees in England, represented a small portion of the ill-fated flux of emigration following Hitler's rise to power in 1933 and the almost immediate enforcement of those racial decrees that had been received by German-Jewish scientists with a mixture of incredulity and dismay; a flux and rising demand of employment which the more far-sighted among the academic institutions in Britain, as well as in the United States, immediately activated to intercept, regarding it as an opportunity to benefit on the front of scientific research and university education, and more in general in terms of the potential infusion of new ideas and new methods (Gavroglu 1995).

In contrast to the aforementioned electrodynamic model of a superconductor by Becker, Heller, and Sauter, the Londons proposed a mathematical model in which the absence of a magnetic field in the superconductor became the *fundamental* law, rather than a *particular* solution of the law of infinite conductivity (the solution obtained by fixing the integration constant H_0 to zero, see Footnote 7 above). In other words, in the Londons' model it was the perfect diamagnetism evidenced by the Meissner effect that uniquely defined the superconducting state rather than its infinite conductivity (London and London 1935). Mathematically, this was done by cleverly replacing the equation of infinite conductivity proposed by Becker and colleagues—according to the Londons, based on a "prejudice which is not tested by

---

[9] This is our translation from the original article, written in German.

experience"—with an equation which allowed as its *only* solution a zero magnetic field in the interior of the superconductor. Differently than the equation of infinite conductivity/modified Ohm's law of Becker and colleagues, the equation proposed by the Londons was static, contained no time derivatives, and related the electrical current j in a superconductor to the magnetic field (through the vector potential $\mathbf{A}$) rather than to the electric field:

$$\mathbf{j} = -\frac{\mathbf{A}}{c\Lambda}.$$

Quite enticing an expression, this was a sort of *magnetic* equivalent of Ohm's law $\mathbf{j} = \frac{\mathbf{E}}{\rho}$ and by formal analogy led the Londons to identifying the Λ with a constant depending on the material, like the electrical resistivity ρ was (Following Becker and colleagues, the Londons were ascribing to Λ the value m/ne^2, with m and e being the electron mass and charge, and n their density). In considering that equation "one is strongly reminded", write the Londons, of the formula proposed a few years earlier by Oskar Klein and Walter Gordon to describe quantum-mechanically the current of an electron with wavefunction ψ in the presence of a magnetic field (Klein 1926; Gordon 1926):

$$\mathbf{j} = \frac{he}{4\pi im}(\psi\nabla\bar{\psi} - \bar{\psi}\nabla\psi) - \frac{e^2}{mc}\psi\bar{\psi}\mathbf{A}.$$

This equation reduces to Londons' equation for a ψ constant in space (since the space derivative $\nabla\psi$ becomes then zero and the product of wavefunctions reduces to the electron density). Unless it was a coincidence, the Londons argued, this fact suggested that the wavefunction of the electrons in the superconducting state ought to have been constant in space across the whole superconductor, and "rigidly" remain so even in the presence of a relatively small magnetic field (proportional to A). How that could occur "atomistically", what would be the microscopic mechanism that generated that macroscopic wavefunction, they only hinted at. The rigidity of the wavefunction against small magnetic fields could be due to a gap in energy separating the electrons in the lowest energy state from the excited state. Such a gap was implicitly contained in the thermodynamic treatment by Gorter and Casimir.[10] But now the Londons speculated that the gap could have been in turn brought about by "some form of interaction" locking the electrons to each other.

With those concluding remarks, the Londons were more than anything opening a research program. For those who believed in the correctness of their intuition, the program laid out the essential questions to tackle the microscopic mechanism of superconductivity. As it would turn out some twenty years later, this was, in fact, the right program.

---

[10] More specifically, the thermodynamic two-fluid model by Gorter and Casimir indicated a reduced density of low-lying electronic states at the Fermi energy.

4.1.2   On a Search for the Microscopic Mechanism

The mathematical model that the Londons proposed, and that Fritz defended from criticism and refined over the few years to come, had a second and larger resurgence when the interest in superconductivity resumed, in the post-war years. Whoever accepted that model and attempted at the hitherto mysterious microscopic mechanism of superconductivity would be guided by a number of elements (Bardeen and Schrieffer 1961). One was that the superconducting state was a thermodynamically stable and reversible state in which the magnetic field was completely expelled. This virtually perfect diamagnetism remained the phenomenological element to be derived microscopically. As the Londons had suggested in their attempt to anticipate a microscopic description, such a diamagnetism could be brought about by a *macroscopic* electronic wavefunction constant across the whole sample and "rigid" against magnetic fields. Since these two features would result from an energy gap at the Fermi energy, the problem boiled down to finding a plausible way to obtain an energy gap from some microscopic arrangement of electrons and ions. The experiments on superconductors that had resumed in the post-war period started to offer the first indirect evidences of such a gap and estimated its value (see Corak et al. 1954, and references to previous experiments therein).[11]

The guiding working hypotheses, constraints, and available experimental clues notwithstanding, not many advances had been made in the understanding of the microscopic mechanism of superconductivity up to 1950, when a breech finally started to open.[12] Part of the reason for this lack of progress had to do with the war situation: the break-out of WWII changed the priorities and subtracted manpower and material resources away from a number of research projects with less immediate applications to channel them on projects of strategic relevance (Hoddeson et al. 1992; Bardeen 2010). We have seen in the previous chapters how, among many others scientists across the continents, Tomonaga and Nambu in Japan as well as David Bohm in America had been respectively involved in research on radars and plasmas (itself related to the atomic bomb project). Since in all that low temperature physics was hardly a priority, interest and experiments in superconductivity resumed only at the end of the '40s and beginning of the '50s.

Other reasons for the stall, specifically with regard to the understanding of the microscopic nature of superconductivity, were no doubt related to the intrinsic challenges that the phenomenon posed from a theoretical standpoint, its elusiveness and subtlety. The most straightforward routes and the hitherto known mechanisms could be ruled out almost from the outset on the basis of energy considerations. Superconductivity took place at extremely low temperatures, which implied that it was

---

[11] These experiments inferred on the existence of the energy gap by measuring the electronic specific heat of a superconductor as a function of temperature. As remarked in, e.g., Serin (1956), these fine and challenging measurements had been made possible thanks to the improvements in the technique of resistance thermometry in the post-war period.

[12] Quite some progress on the understanding of the superconducting transition in the vicinity of the transition temperature had been made with the 1950 phenomenological treatment of Vitaly Ginzburg and Lev Landau (republished in English in Ginzburg and Landau 2009). The treatment, however, did not shed much light on the microscopic nature of the superconducting ground state.

an extremely low-energy process. As confirmed by the measurements on the energy gap, the process involved changes in the energy per electron as low as 10^{-8} eV.[13] Indeed, that was a minute energy compared to that of other known phenomena in solids, and was below the accuracy that the available approximation methods used in many-body physics could reach. This posed both a conceptual and a mathematical challenge. To begin with, the question of what the mechanism was required imagination; then, once hypothesised, one would face the challenge of numerically proving its very plausibility.

With regard to the physical mechanism—as a nudge to the fantasy—the waters began to clear as the results from a simple experiment, conducted independently by two research groups, were issued in May 1950 (Maxwell 1950; Reynolds et al. 1950). Rather surprisingly, the experiments showed that two superconductors, equivalent in all respect but for the *nuclear* masses of the atoms of the ionic lattice, became superconducting at different temperatures.[14] The isotopic mass would not have been an important parameter unless the motion of the ions was directly involved in superconductivity, and thus this fact strongly suggested that the latter must have arisen from some sort of interaction between the electrons and the motion of the ionic lattice. This result, christened the "isotope effect", came in fact as an indirect confirmation of an idea on which the physicist Herbert Fröhlich was building his model of superconductivity, and happily encouraged his intuition (Fröhlich 1950; Bardeen 2010). Fröhlich had been modelling what happens to an electron as it moves about in a crystal displacing the ions with its electric field. In particular, he had investigated if and for which conditions (velocity of the electron, frequency and distribution of the vibrational modes of the crystal) such interactions resulted in a lowering of the energy of the electron as compared to when the electron propagated in vacuum. Assuming this energy arose as a consequence of the reaction of the lattice deformation/polarization to the field produced by the electron itself, Fröhlich performed this estimation by adapting the concept of self-energy from quantum field theory. The perturbative calculations indicated that in certain conditions such interactions between the electrons and the quanta of lattice vibrations (later called phonons) resulted in a *negative* self-energy. This implied an *attractive* interaction of the electron with itself and thus a lowering of its energy. In working that out, Fröhlich had the idea that analogous attractive interactions in metals could be responsible for the lowering of the ground state energy and the emergence of an energy gap at the Fermi energy.

According to this logic, the stronger the electron-phonon interactions, the easier and more likely it would be for a metal to become superconducting. As those same interactions between electrons and phonons were held responsible for resistivity in metals, this also meant that bad conductors were most likely good superconductors, and vice versa. The intuition seemed to be roughly consistent with the results of the

---

[13] To give an idea of the smallness of this energy value to the unacquainted reader, we may notice that the mean thermal fluctuation energy at room temperature is 2.5×10^{-2} eV, which is 10^6 times larger than that value; and the nucleon-nucleon binding energy is roughly 8 MeV, that is, 10^{14} times larger.

[14] In particular, the transition temperature was shown to depend inversely on the square root of the isotopic mass $T_c \propto \sqrt{1/M}$.

experiments on which metals did and which did not become superconductors, and at which transition temperatures they did so. "It is not accidental", Fröhlich argued, "that very good conductors [i.e., gold and copper] do not become superconductors, for the required relatively strong interaction between electrons and lattice vibrations gives rise to a large normal resistivity [i.e., the resistivity of the non-superconducting state]" (Fröhlich 1950). This correct reasoning definitely showed the inconsistency of the previously held conception of the superconducting state as an ideal metallic state.

Advancing this hypothetical scheme, Fröhlich had in effect grasped an important aspect of superconductivity: the fact that it was most probably the *dynamics* of the lattice that was involved in superconductivity; and, more specifically, that the phonons could be responsible for some sort of attractive interaction, which in turn could be the cause of the energy gap. On the other hand, his model was incorrect as to the exact way such interactions came about and their effects were computed; and indeed it did not manage to reproduce the gap.[15] In retrospect, these shortcomings were a consequence of the way in which Fröhlich had approached the problem from the mathematical standpoint, and the fact that he had neglected the Coulomb interaction.

Starting from the Hamiltonian featuring the electron and phonon fields and the electron-phonon interaction, as is customarily done in field theory, Fröhlich had regarded the conduction electrons as sources of phonons and tried to obtain the attractive interaction, and thus derive the gap, purely from the *perturbative* self-interaction of the electrons in the phonon field. In that procedure, as it started to become clear with the subsequent works of Robert Schafroth (1951) and Fröhlich (1954) himself, the limitation was, to begin with, inherent to the mathematical tools that were used— the perturbative scheme of quantum field theory. As proven by the two physicists, the energy gap could never have been obtained by means of perturbative self-energy calculation. This was because the gap was found to depend in a non-analytic way on the electron-phonon coupling strength, and the perturbative expansion in powers of that coupling would automatically yield a vanishing gap regardless of the order of perturbation and the value of that coupling. In short, the phenomenon of superconductivity was transparent to the perturbative methods customarily employed in quantum field theory, thus imposing the use of non-perturbative methods to even hope to explain it. On top of this mathematical limitation, there was another limitation. Using the standard self-energy, the procedure adopted by Fröhlich in his 1950 model could in no way include the interactions of the electrons with the other electrons mediated by the exchange of virtual phonons, and thereby had been excluding many-body effects. As was understood in the coming years, superconductivity emerged precisely out of those many-body effects.

---

[15] For those keen on the details, the inadequacy was two-fold. The energy due to the self-interactions (diagonal terms of interaction Hamiltonian) that Fröhlich's and affine models were obtaining was much too high compared to the superconductive energy gap. And, equally important, that self-energy was found actually both in the normal and in the superconductive state—so as to cancel each other out when taking the energy difference—and thus could not have been responsible for the emergence of the superconducting state (see, e.g., Bardeen 1956).

The failure to reproduce the main feature of superconductivity with the tools of perturbative quantum field theory, at any order in perturbation, evidenced the fact that the basic mistake was the attempt to treat an intrinsically collective phenomenon as a standard single-particle problem. On the physical level, while the Coulomb repulsive interaction between electrons as the only source of superconductivity had been excluded on the basis of energy considerations, it might have played a role *in combination* with the interactions mediated by the phonons.[16] As we see in the following, superconductivity would emerge indeed out of a subtle balance between these two interactions.

4.1.3 Bardeen, Pines, Cooper, Schrieffer, or the Resolution

All this negative and positive evidence about superconductivity, gradually accumulated from 1950 onward as a result of experimental and theoretical efforts, began to converge in 1955 and resolved into a coherent picture two years later, with the so-called Bardeen-Cooper-Schrieffer (BCS) model. The process was led by John Bardeen, the mastermind who put together the results of the collaborations with his younger students David Pines, Robert Schrieffer, and Leon Cooper. In his late forties at the time, an absolutely brilliant mind well-versed in electrical engineering as well as theoretical physics and a man of proverbial taciturnity, modesty and understatement, Bardeen had hitherto applied himself to a variety of problems in the field of solid-state physics of theoretical and practical interest.[17] Among other problems, he had addressed the many-body interactions between electrons and phonons in metals (Bardeen 1937), investigated the hypothesis of static lattice displacement as a possible cause of the superconducting gap (Bardeen 1951 and previous references therein), and had performed the work on semiconductors which led him to the momentous invention of the transistor in 1947 (Daitch and Hoddeson 2002). Having learned of the isotope effect directly from one of its discovers, Bardeen had rekindled his previous interests in the topic and worked out a model for superconductors along the line of Fröhlich's (Bardeen 1972). In collaboration with the latter—who in late 1950 was a visiting scientist at the Bell Laboratories (Bardeen 2010)—and a few of the physicists who were determined in those years to understand superconductivity from a microscopic perspective, Bardeen had systematically contributed to, and critically assessed, the various results and routes to superconductivity that we have detailed above (Bardeen 1956). In this endeavour, aware of the theoretical challenges that the problem posed, he had opted for what he called an "inductive" approach. He summed this up in 1955 in the chapter reviewing the progress in the theory of

---

[16] As a reminder, the Coulomb repulsive interaction in metals is in the order of 1 eV per electron. This energy scale is significantly higher than the 10^{-8} eV per electron typical of superconductivity.

[17] Biographical sketches of the person and the scientist, from more anectodic to more properly biographical, are given by Laughlin (2006) and Pines (2009). A well-rounded portrait is given by Hoddeson (2002).

superconductivity in Springer's *Handbuch der Physik* on low-temperature physics (Bardeen 1956)[18]:

> Anything approaching a rigorous deduction of superconductivity from the basic equations of quantum theory is a truly formidable task. The energy difference between normal and superconducting phases at the absolute zero is only of the order of 10^{-8} eV per atom. This is far smaller than errors involved in the most exacting calculations of the energy of either phase. One must neglect terms or make approximations which introduce errors which are many orders of magnitude larger than the small energy difference one is looking for. One can only hope to isolate the physically significant factors which distinguish the two phases. For this, considerable reliance must be placed on experimental findings and the inductive approach.

As was becoming clear to him, the physically significant factor in the superconducting phase hid in the equilibrium between the screening time-scales of the electron system and the ion system: how fast charge perturbations travelled through the system of electrons relative to how fast the ionic lattice reacted to screen them (Bardeen and Schrieffer 1961). Intuitively speaking, the idea can be visualised by imagining a mobile electron in a metal surrounded by other such electrons as well as the deformable positively charged ionic lattice. Upon a change in the electron's linear momentum—transitioning from a momentum state k to a state $k + q$—an electron in the metal locally causes a small perturbation of the negative charge density. Owing to the strong electron-phonon coupling that had been identified as characteristic of superconductors, such perturbation in the charge distribution (and electric field) generates a phonon. The phonon, amounting to a deformation of the ionic lattice, gives rise to a modification of the positive charge density surrounding the electron. Depending on the relative speed of these electronic and ionic screening processes, there will transiently be an excess or a deficit of positive charge around the electron. If the energy/frequency of the phonon is high enough, as is the case in superconductors, there will be an excess of positive charge around the electron. The electron will thus appear to a second electron in the surrounding as having a net positive charge, with the result that the two electrons will be attracted to each other. Effectively, this is a process in which a virtual phonon mediates an electron-electron attractive interaction which overcomes the Coulomb repulsion.

The subtlety of this mechanism and its entirely dynamical nature were the reason why it had hitherto eluded explanation. The range of states of energy and momentum of the electrons and the phonons for which these complex dynamics—resulting in the net attractive potential—took place was found to be very narrow, and involved changes in electronic energy of, finally, the sought-for order of magnitude. That was why those relatively few processes, while washed out at high temperatures and contributing insignificantly to the normal state, would have occurred only at low enough temperatures and determined the transition to the superconducting phase. As argued by Bardeen, at such temperatures, the effective electron-electron interaction between those few very specific electronic states was so large that it could not be seen by the previous perturbative treatments, and required instead a separate non-perturbative treatment (Bardeen 1956).

---

[18] Bardeen's chapter in the *Handbuch der Physik* was written in 1955, but published the year later.

This picture sketched in 1955 provided the base framework that two years later would eventually result in the successful microscopic model of superconductivity. Far from intuitive, Bardeen had arrived at the picture with the central collaboration of David Pines by distilling, as it were, the one key interaction from the complicated mass of interactions in the solid that possibly gave rise to the superconducting state.[19] Starting from an Hamiltonian which factored in all these interactions, Bardeen and Pines managed to extract, using the appropriate canonical transformation, a single interaction potential between electrons which effectively included the relevant contributions and factored out the supposedly non-relevant ones (Bardeen and Pines 1955; Bardeen 1956). In this master work of heuristic inductive reasoning that shines through Bardeen's long chapter in the *Handbuch*, the operation was carried out using a pondered mixture of simplifications, mathematical manipulations, and comparison with the precedent attempts, weighing, with more intuitive or mathematical/dimensional arguments, the probability that each of them would play a part in superconductors.

In extracting the effective electron-electron interaction potential, Bardeen and Pines had transformed the problem of the many-body interactions between electrons and phonons in all its complexity into the *simpler* problem of treating a gas of electrons with *attractive* interactions, to be solved with non-perturbative methods (Bardeen 1956). How this could give rise to superconductivity though was yet to be demonstrated; and simpler certainly did not mean easy. In that case, mathematically speaking, the problem was still "an extraordinarily difficult problem" as Leon Cooper remembered realising at the end of 1955 when actually setting out to solve it (Cooper 2011). Conscious of the challenge and of the need for resources, Bardeen had hunted for brilliant brains that were ready to help—beside his graduate student Robert Schrieffer who had superconductivity as his doctoral thesis problem—and found one in the young post-doctoral fellow Leon Cooper. Cooper had hitherto studied and worked on nuclear and particle physics, and as such was barely aware of the existence of superconductivity (Cooper 2011). Having acquainted himself with the status quo of superconductivity, and after a number of failed attempts in various directions, Cooper identified the simplest possible configuration to work with that preserved the salient features of the problem. In roughly half a year of mathematical work, he reached the key result. The gas of electrons in the metal became unstable upon switching on an attractive interaction V between the electrons at the Fermi energy, and led to the formation of pairs of electrons of opposite spins and momenta.[20] As a result, the energy of these electrons lowered by an amount $\sim \hbar\omega e^{-1/V}$ with respect to the energy in the metallic state, and led to a gap in the electronic energy spectrum. The gap separated the ground state formed by such pairs and the conduction electrons in

---

[19] The total Hamiltonian that Bardeen and Pines start from included the electrons, the ions of the underlying lattice, the phonons, and all the coupled dynamics brought about by the Coulomb forces and the electron-phonon interaction (see Sect. 37 "Derivation of the Hamiltonian" in Bardeen (1956)).

[20] These were the states with zero total momentum whose weight was practically zero in the normal state, but which were subject to a strong coupling in the superconducting phase and that had eluded the perturbative treatment (Bardeen 1956).

the standard metallic state: its numerical value and dependence on the energy of the available phonons ($\hbar\omega$) corresponded to the form and value of the energy gap which had been sought for, and indeed turned out to be it.

On the basis of this result, towards the end of 1956, Schrieffer arrived at the mathematical form for the ground state formed by many such pairs (Warnow and Williams 1974). As he recounts, the right solution came to him one day on a subway ride in New York. That wavefunction, which embedded the pairs found by Cooper, described an ordered state in which the electrons of opposite spin and momenta within a small region around the Fermi energy formed pairs and "condensed" into the same zero net momentum state. It was the breaking off of one such pair, setting an individual electron free from the superconducting cooperative ground state into the normal metallic phase, that cost on average an amount of energy equal to the gap. The fact that each of those electrons was coordinated with its counterpart in spin and momentum at "large" distances across the superconductor endowed the ground state wavefunction with that "rigidity" that the Londons had suggested, and explained from a microscopic point of view the reason for the vanishing of the electrical resistance and perfect diamagnetism that had lingered on unexplained for about fifty years. The detailed calculations that the three-man team performed at a tight pace over the next five months quantitatively showed that this was the case. This effort led in July 1957 to the thirty-page long *Physical Review* article plainly entitled "Theory of Superconductivity" (Bardeen et al. 1957). This so-named BCS model (after the authors) reproduced with quite some accuracy the thermodynamic, electrodynamic and transport and non-equilibrium properties of the superconductors hitherto experimentally measured. Cooper later remembered the thrill of the moment (Cooper 2011):

> The next five months were a period of the most concentrated, intense and incredibly fruitful work I have experienced. We divided our efforts: Bob [Schrieffer] would focus on thermodynamic properties, I would focus on electro-dynamic properties, with John [Bardeen] focusing on transport and non-equilibrium properties. New results appeared almost every day. John's vast experience came to the fore. All of the calculations done previously with a normal metal (these John had in his head) could be turned on this new theory of superconductivity if one put in the appropriate matrix elements.

With this solid quantitative model and qualitative explanations, superconductivity was completely understood. And yet, there was one element of principle that blemished the coherence and completeness of the scheme: the issue of electric charge conservation, or gauge invariance, of the superconducting ground state.[21] The wavefunction of the superconducting ground state was not invariant under that capital symmetry of electromagnetism which is intimately connected to the conservation of electric charge and, as a consequence, the calculation of the Meissner effect formally depended on the choice of gauge. Since the effective electron-electron interaction that had been derived by Bardeen and Pines (and was the core of the BCS model) contained a momentum-dependent cutoff which rendered it manifestly non-gauge

---

[21] We beg the reader who is puzzled before these notions to wait for the next section for a clearer explanation.

invariant, there remained a fundamental doubt about whether the lack of gauge invariance of the Meissner effect was an artefact of the BCS treatment, or rather somehow inherent to superconductivity. As pointed out in a footnote added in proof by Bardeen (note 36a in Bardeen et al. 1957) and shown more explicitly later (Anderson 1958a), the processes excluded by the momentum-dependent cutoff could not be held responsible for the lack of gauge invariance. In this sense, the BCS model was *approximately* invariant. These elements notwithstanding, the lack of *exact* gauge invariance cast a shadow on the soundness of the electromagnetic properties of superconductors that were predicted by the model: the question was how a theory about an object that does not abide by one basic principle of electromagnetism could be right about the system's interaction with the electromagnetic field.

To be sure, in the face of the overwhelming quantitative success that the model scored, that issue of principle—of which the team had been aware from the start, according to Cooper—had been deliberately given less priority and set aside earlier on (Cooper 2011). That was decided by Bardeen who had preferred to concentrate the resources of his team on drawing the numerical conclusions, having estimated that the issue indeed made a negligible difference for the latter (Bardeen 1957). The fact remained that, in the aftermath of the theory of Bardeen, Cooper, and Schrieffer, the issue of gauge invariance was still formally unsolved. Various physicists, including Nambu, were disturbed by the issue and set out to solve it one way or another. Building on an elegant reformulation of the BCS theory given by Bogolyubov, the reformulation by Nambu will definitively settle the question of gauge invariance. At the same time, in doing so, it will also unknowingly open a whole new dimension of theoretical possibilities.

4.1.4 Bogolyubov's "New Method", or the Reformulation of Superconductivity

The simplification that Bardeen had very efficaciously employed, consisting in replacing a complicated interaction between electrons and phonons by an effective and simpler one between electrons, isolated the core aspects of superconductivity and enabled him to account for the gap and the ground state of the superconductor. However, as we said, that operation made the BCS Hamiltonian manifestly non-gauge invariant, and eliminated the phonons from the picture. In that sense, the effective BCS Hamiltonian might have hidden a part of the spectrum of phenomena that were contained in the complete *gauge-invariant* Hamiltonian it had been derived from. Restarting from the Fröhlich Hamiltonian, which was gauge-invariant and included the phonon field, a few months after the BCS model Nicolay Nicolayevic Bogolyubov offered a reformulation which gave a more complete picture of superconductivity.

Of the same generation as Bardeen, Bogolyubov was not dissimilar to the younger Nambu in coupling eclectic interests with a quite outstanding mathematical deftness. A mathematical physicist by education, Bogolyubov employed his passion for, and abilities in, mathematical methods to travel across a wide range of problems and contexts in physics and mathematics, as it shines through his scientific production. Such an attitude was in fact not that unusual in the Soviet Union, where physicists in most cases were versed in a number of areas comparatively larger than in Europe and the United States. Leaving the question for an elaboration elsewhere, here we just suggest that this was, at least partly, due to the strong linkage enforced in the Soviet Union between science & mathematics and technology, fundamental and applied research (see, e.g., Kojevnikov et al. 2008). These differences of context considered, Bogolyubov's attitude remained exceptional for the breadth and depth of his contribution. By way of invention, generalisation, or transfer of mathematical techniques and approximation methods, he attacked problems in classical and quantum statistical mechanics, many-body problems in solid-state physics, quantum field theory, and dynamical systems, shifting between them in his educational and professional years.[22]

As one of such applications, in 1947 Bogolyubov had approached the problem of superfluidity with the conceptual and mathematical resources of quantum field theory (the method of second quantisation) with the aim of modelling the excitations of the superfluid (Bogolyubov 1947). By way of a transformation of coordinates— now named after him—he had shown that such excitations could be "described as [...] certain "quasi-particles" [...] which cannot be identified with the individual molecules" of the superfluid. A decade later, returning to low-temperature phenomena as the microscopic picture of superconductivity was approaching a solution, Bogolyubov carried out an analogous operation with superconductors (Bogolyubov 1958; Bogolyubov et al. 1958). The operation, along with its physical interpretation, formed the base of his reformulation of the picture that Bardeen, Cooper and Schrieffer had just proposed. Mathematically, this consisted of the canonical transformation of the electron field operators featured in Fröhlich's Hamiltonian into a new pair of fermionic operators. The new set of fermions, called "quasi-fermions" or "quasiparticles", were composed partly of an electron and partly of a hole with opposite momentum and spin. The corresponding ground state/"vacuum" of the superconductor was thought of as formed by such non-interacting quasiparticles. The Bogolyubov transformation was designed so as to incorporate in the operators all the non-perturbative divergent contributions contained in the interaction term and returned a new Hamiltonian featuring quasiparticles interacting only via a small residual interaction.[23] The transformation was effectively defining and describing

---

[22] So the committee of the Daniel Heinemann Prize for Mathematical Physics justifies the award given to Bogolyubov in 1966: "For several outstanding achievements in bringing the resources of modern mathematics to bear upon fundamental problems of physics [...]". The list of prize recipients and related motivations can be found on the webpage https://www.aps.org/programs/honors/prizes.

[23] The coefficients of the superposition, defining the proportion of the electron and hole components making up the quasi-particle mixture (the new fermion operators), were appropriately chosen so that the divergent terms $\sim 1/q$ (where q is the momentum of the phonon exchanged between two electrons) corresponding to the creation of quasiparticles out of the vacuum vanish. In this way, the terms of the perturbation expansion which would not be treatable with perturbation theory were eliminated in this new representation in which quasi-particles replaced electrons.

in the most convenient way quasiparticles that could be excited/generated from the ground state/vacuum state of the superconductor.

In analogy to what occurred in the superfluid, the quasiparticles found by Bogolyubov differed essentially from the standard electrons making up the super-conducting material and had distinctive properties (Bogolyubov et al. 1958). They corresponded to a "kind of superposition of an electron and a hole", where the pro-portion of the electron component (negative charge) and hole component (positive charge) depended on the proportion between the quasiparticle's kinetic energy and the energy gap. A central implication of this was that the electric charge of these quasiparticles was no longer fixed and negative, but varied proportionally to the particle's velocity; and was thus no longer a conserved property. It was these quasi-particles, rather than electrons as BCS model predicted, that would appear as the elementary excitations of the superconductor (when providing to the latter an energy larger than the energy gap). These quasiparticles tended to electrons only for kinetic energies (and momenta) much larger than the energy gap, recovering in this limit the particular case predicted by the BCS theory.[24] Besides existing as free (above the gap), the fermionic quasiparticles in Bogolyubov's framework also formed bound pairs as a result of the residual phonon interaction. These pairs, bosons, represented the quanta of the collective excitations of the superconductor whose energy lay below the gap.

The quasiparticles with special properties that were uncovered by Bogolyubov's reformulation settled a fundamental question that had remained open in BCS. Since the quasiparticles emerged from a gauge-invariant electron-phonon interaction, they revealed that the violation of charge conservation was not a defect of the simplified BCS treatment, but something essential to superconductivity, that is, connected to the very presence of a non-zero gap. The symmetric electron-phonon interaction that was at work in superconductors somehow gave rise to a set of non-symmetric/non charge-conserving manifestations. In this way, the reformulation given by Bogolyubov had the merit of exposing more transparently the anomaly, and peculiarity, of the super-conducting state. This fact, together with the physical interpretation of the excitations that it put forward, were the elements that opened the way for Nambu.

4.1.5 Discovering by Way of Translating: Nambu and the "Mathematical Structure" of Superconductivity

The microscopic description of superconductivity given by Bardeen, Cooper, and Schrieffer and the subsequent reformulation of Bogolyubov stimulated a particular interest in Nambu. In fact, Nambu had begun to be more closely interested in the

---

[24] This can be seen from the quasiparticle's wavefunction (see equations in Sect. 4.2.2). For kinetic energies much smaller than the energy gap, the quasiparticle is an equal superposition of an electron and a hole, while it effectively becomes a free electron (hole) for high positive (negative) linear momenta.

phenomenon at the beginning of 1956. From the first note dated January 10 of that year, a notebook on electron gas and theory of superconductivity (Nambu 1956) had been intermittently filled along with a few other threads he was upon in parallel (mostly mathematical structure of dispersion theory and Green's functions, and weak processes). This work on superconductivity is sometimes more speculative and sometimes consists of the re-working of the results of others in different approximations and limits with the formalism of quantum field theory. In early 1957, on the occasion of a seminar given at the University of Chicago where Nambu had relocated, Schrieffer had exposed his work in progress and the hypothesis he had formulated for the wavefunction of the superconducting ground state (Ashrafi 2004; Nambu 2016). At once enticed and puzzled by the problem of the violation of charge conservation/lack of gauge invariance—as well as Schrieffer's nonchalant acceptance of it—from then on Nambu began his more systematic work on superconductivity. That "unsettled problem" seems to have lingered on in Nambu's background until the final model issued by Bardeen, Cooper, and Schrieffer and its reformulation by Bogolyubov prompted him to take it up more decisively (Nambu 2010). Given the immediate resonance that the solution of the mystery of superconductivity offered by Bardeen, Cooper, and Schrieffer had had in the scientific community (after about fifty years superconductivity was finally understood!), it should not be a surprise that in its aftermath several theoretical physicists were, like Nambu, equally discomforted by that problem of principle blemishing the BCS solution, and set to work on it. The list of them, each proposing their own possible solution, is fairly long: it included Bardeen, Schrieffer, and Pines themselves, Nambu's head of the group Gregor Wentzel, as well as other solid-state and statistical physicists.[25] Together with Bogolyubov, Nambu stands out there as the only particle physicists—as far as his eclectic interests permit such a definition. In hindsight, only Nambu's turned out to be a real final solution. The others had solved the problem, but "locally" and thus failed to identify what the lack of gauge invariance implied at bottom. This had to do with the fact that Nambu, though motivated by the same problem, arrived at its solution by way of a complete reformulation of superconductivity (Nambu 1960e), a picture which reproduced the results of BCS and Bogolyubov and, at the same, established the links that were missing in theirs.

The reformulation that Nambu had in mind would be based on the collective description that he had developed with Kinoshita for plasmas and nuclei, and that, unsuccessful in treating the latter, had sat silent and practically uncited ever since (see Sect. 3.2.2). Adapting that description to superconductors, Nambu planned to address the problem of gauge invariance with the mathematical tools and theorems already available in quantum electrodynamics. In effect, that formalism provided a quite general field-theoretical formulation of the many-body problem—built on the Feynman and Dyson's quantum electrodynamics and the Hartree-Fock method—through which one could also potentially address the specific question which had meanwhile arisen in the context of superconductivity: the question of modelling the non-perturbative attractive interactions between many electrons mediated by a phonon field. The possibility of covering superconductivity was granted by the fact

---

[25] A quite complete list is provided in the introductory section of Nambu's article (Nambu 1960e).

that, in contrast with the formalism of Bohm and Pines, Nambu's formalism structurally treated fermions and bosons on equal footing. This generality and flexibility, which only now turned into a major formal advantage, had been attained by introducing the concept of self-consistent self-energy (see Sect. 3.2.2). Through the latter, one could include the potential experienced on average by a single fermion or boson immersed in the medium in the respective self-energies, and obtain new renormalised particles and their energy spectrum. Instead of undergoing (non-linear) four-particle scattering, those particles could be approximately described as moving in an effective (linear) potential that included the effects of that scattering. In the case of plasmas, massless photons were transformed into massive photons, while electrons were left essentially unchanged in their properties. In the case of the superconductor it was the electrons that should have been radically transformed, via the action of an effective potential (supposedly more complex than that of the plasmas), so as to give rise to the quasiparticles found by Bogolyubov—superpositions of electrons and holes.

Obtaining in his formalism Bogolyubov's quasiparticles from electrons is the first task to which Nambu sets himself. This consists in working out the corresponding self-consistent self-energy/Hartree potential that the original electrons are endowed with/subject to as a consequence of the electron-electron attractive interaction. In order to do that, Nambu writes the generic quasiparticle as a two-component wavefunction $\Psi = \begin{pmatrix} \psi_e \\ \psi_h \end{pmatrix}$, where the first component represents the electron and the second the hole. He then writes the quasiparticle self-energy $\Sigma(p)$ as a sum of two self-consistent potentials, a diagonal, $\chi(p)$, and a non-diagonal one, $\phi(p)$ (see Eq. 2.14 in Nambu 1960e):

$$\Sigma = \chi \begin{pmatrix} 1 & 0 \\ 0 & -1 \end{pmatrix} + \phi \begin{pmatrix} 0 & 1 \\ 1 & 0 \end{pmatrix} \tag{4.1}$$

If the diagonal term yields an electron and a hole modified merely in their respective kinetic energies, ϵ, the presence of the non-diagonal self-energy term couples the electron with a hole of opposite momentum and spin. In this way, that term changes the electron into a quasiparticle that is no longer an eigenstate of charge and whose charge/gauge symmetry properties are radically transformed. The possible energy states of the resulting quasiparticle are $E = \pm\sqrt{(\epsilon + \chi)^2 + \phi^2}$. This expression bears a formal resemblance, Nambu writes down in a note at the beginning of 1959 (Nambu 1959), to the energy states of a Dirac fermion $E = \pm\sqrt{\epsilon^2 + m^2}$, upon replacing the mass term m with the Hartree potential ϕ. When looked at through Dirac's hole picture, which Nambu had used as the concrete base of his formalism (see Sect. 3.2.1), the resemblance has a straightforward physical interpretation. The ground state/vacuum state of the superconductor is a state in which all the negative energy quasiparticle states are occupied, and a quasiparticle can be excited from the negative ($E < E_F$) to a positive ($E > E_F$) energy state only by providing an energy to the superconductor $E \geq 2\phi$.

In Nambu's framework, the two potentials χ and ϕ to which the electron is subject are determined by solving two self-consistent coupled equations $\chi = \chi(\chi, \phi)$ and $\phi = \phi(\phi, \chi)$. From these equations, it emerges with particular clarity that, while a non-zero value for the diagonal component χ of the self-energy can result from a perturbative expansion, a finite non-trivial solution for the non-diagonal term ($\phi \neq 0$) can be derived only in the non-perturbative treatment. This special solution is calculated by Nambu to be nothing but the familiar energy gap $\phi \sim \hbar\omega \exp(-1/V)$, found by Bardeen, Cooper, and Schrieffer. As is clear from the mathematical form of the self-consistent equation for ϕ, such a non-trivial solution could not follow from a perturbative expansion at whatever order as there is no inhomogeneous term $\phi_0 \neq 0$ to start with (see Eqs. 2.18 and 2.19 in Nambu 1960e). In other words, the perturbative expansion is seen to provide only one particular solution $\phi = 0$ (higher in energy and metastable, where no gap is present) failing to yield the more general one.

Albeit reached from a different route, the quasiparticle obtained in Nambu's reformulation is analogous to Bogolyubov and equally violates the conservation of electric charge. When in the presence of an external electromagnetic field, the current of quasiparticles, in contrast to that of a regular charged particle, does not satisfy the continuity equation. This implies that electric charge can be generated out of nowhere and dissolve away. As already shown by Bogolyubov, the very existence of the energy gap—a non-zero self-consistent self-energy term ϕ in Nambu's framework—which brings about the quasiparticles inevitably brings as well a violation of charge conservation/gauge invariance.

If all the results obtained by Nambu until here are, in fact, a mere restatement of Bogolyubov's formulation, the mathematical structure of his formalism grants a longer sight. Exploiting the tools built in quantum electrodynamics, and in particular guided by well-established theorems about gauge invariance therein, he can show transparently what are the conditions to preserve gauge invariance, and that the latter is in fact preserved when we consider that quasiparticles are "dressed" by the collective modes. The idea is essentially to treat the quasiparticle as the new bare particle of the theory, and consider the phonon interaction on it as one considers the photon interaction on the bare electron. Calculating with a similar procedure as in quantum electrodynamics the "radiative corrections" to the bare quasiparticle due to the *phonon* interaction, Nambu is able to derive a modified ("dressed") charge-current operator for the quasiparticle. This yields an electric current for the dressed quasiparticle, J, given by the current of the bare quasiparticle, J_{QP}, *plus* an additional term. If J_{QP} alone did not fulfil the continuity equation, now J does. More specifically:

$$J = J_{QP} - 2\phi\, e\, \partial_\mu f$$

In this expression, the additional term $-2\phi\, e\, \partial_\mu f$ physically represents the electric current carried by the collective longitudinal modes, f. These collective modes/quanta make up the "cloud" of virtual excitations generated by the bare quasiparticle when it is accelerated, and compensate for that portion of charge that in Bogolyubov's picture seemed to be created from nowhere and dissolve.[26] These

---

[26] Nambu will speak of the charge being "exchanged between the core and the cloud" of Bogolyubov's quasiparticle when accelerated (see Eqs. 7.11–12 in Nambu 1960e).

modes correspond exactly to the longitudinal collective modes that had been found by Bogolyubov. However, in Nambu's framework they are found to perform the *essential* function of restoring the charge conservation/gauge invariance that seemed violated as a consequence of the presence of the energy gap. In other words, the presence of quasiparticles *and* the requirement of the gauge invariance—seemingly incompatible with each other—are shown to be compatible with each other by automatically implying the existence of the longitudinal collective modes.[27]

In solving the problem of gauge invariance in this way, Nambu's reformulation exposes a deeper and underlying logical connection between the manifestations of the superconductor (energy gap, quasiparticles, collective excitations) and the symmetry properties of the equation from which these manifestations descend. *Both* quasiparticle excitations and collective excitations are necessary consequences of the superconducting ground state and the energy gap ϕ. These are non-gauge-symmetric outcomes when considered separately, but have the same symmetry of the dynamical equation when considered together. Said in other terms, if the gauge-symmetric phonon-mediated attractive interaction leads to the formation of an energy gap ϕ, and thereby quasiparticles that violate the gauge symmetry/charge conservation, at the same time it is responsible for the emergence of longitudinal collective modes which restore it.

By the time Nambu reaches this solution, as we already mentioned, the problem of gauge invariance had in fact already been addressed and actually solved by others. These had recognised the role that longitudinal collective modes had in exactly cancelling out external longitudinal fields. Unlike the others, however, Nambu's solution is not just neutralising, as it were, the *negative* result, the issue troubling the model of Bardeen, Cooper, and Schrieffer. The logical connection between the asymmetry of the manifestations of the superconductor and the symmetry of its equation that he exposes unearths a general theoretical possibility which will be used *positively* in the realm of particle physics, as we will see. By positive use we mean its use as a mechanism one can take advantage of and that one seeks in order to generate particles' properties dynamically and diversity (asymmetry) out of uniformity (symmetry). In this way Nambu's reformulation transforms the problem to fix in superconductivity into a problem to deliberately invite in elementary particle physics, as there it is a resource.

If this constitutes the more epistemological element preparing the transfer to particle physics, there is also one element on the technical/perspectival side. This has to do with the fact that in Nambu's reformulation the energy gap is incorporated as a part of the electron self-energy—and thus is regarded as a property of the electron. The logical connection that Nambu exposes entails then that bare fermions, if made to interact in a certain way, can give rise to dressed fermions that have an entirely dynamical property—a property completely absent in the bare fermions—which lowers the original symmetry. In this sense, Nambu's particle physics perspective on

---

[27] One of the consequences was the gauge-invariance of the Meissner effect: no current appeared in the theory when one inserted a *longitudinal* static magnetic field ($q \to 0$) in the formula relating the current of a superconductor with the external electromagnetic field $J_i(q) = \sum_{ij} K_{ij} A_i(q)$.

superconductivity is already adapted to its future application to particle physics. The analogical correspondence between the ground state of the superconductor and the vacuum of particle physics, quasiparticles and elementary particles is potential and just waits for a trigger to go off.

These are the subtle but essential premises for the transposition of a phenomenon local to superconductivity to the realm of particle physics, where it becomes a generative theoretical possibility.

The reasoning we have reconstructed here is fleshed out in two dense notebooks (with the titles "Superconductivity I and II") filled up in May 1959, at one point featuring an "everything is clear!". Structured in a more linear form, this is essentially the material that will make up the long article entitled "Quasi-particles and Gauge Invariance in the Theory of Superconductivity" and submitted by Nambu to *Physical Review* on July 23rd 1959 (Nambu 1960e). As we will see in a moment, these dates acquire an historical relevance here when put in correspondence with a passing comment of Nambu made at the Rochester Conference on High-energy Physics in the summer of that same year.

4.2 The Universe as a Superconducting Solid

If the complex of manifestations that Nambu just discovered was hitherto limited to the superconductor, it was equally true that their characterization already appeared in a sufficiently general form. This was owed, in turn, to the structure of the formulation itself, which was constructed as a generalisation of quantum electrodynamics. While carrying out his reformulation of superconductivity, Nambu already prefigured, if only from a conceptual point of view, a potential application to another domain of physics, the more fundamental domain of elementary particle physics. These thoughts are publicly voiced at the 9th International Rochester Conference on High-Energy Physics which takes place between 15–25 July 1959, that year to be held in Kiev, Ukraine, and to which conference Nambu had left immediately after submitting the article on gauge invariance in superconductivity. At the conference Nambu does not officially present anything, but the trace he leaves in the form of a brief comment dropped in one of the open discussion sessions betrays most eloquently what his mind is incubating at that moment. The session brings the generic heading "theoretical investigations" and Nambu's intervention, indirectly addressing a comment made by Heisenberg on the latter's unified field theory of elementary particles, goes as follows (Nambu 1960b):

> I would like to call attention to an analogy between the problem of γ_5-invariance [i.e., the chirality invariance] in field theory and that of gauge invariance in superconductivity as formulated by Bogolyubov. In this analogy the mass of an observed particle corresponds to the energy gap in a superconductive state. The Bardeen-Bogolyubov description of superconductivity is not gauge invariant because of the energy gap. But several people have succeeded in interpreting the Bardeen-Bogolyubov theory in a gauge invariant way. In this way one could also handle the γ_5-invariance. Then, for example, one easily concludes the existence

of a bound nucleon-antinucleon pair in a pseudoscalar state which might be identified as the pion state. So there seems to be a possibility of learning something about the elementary particles by studying superconductivity.

In recalling the problem of chiral invariance, Nambu refers here to an open problem in the field of particle physics to which the mechanism which he just discovered at work in superconductors seemed to offer, by analogy, a solution. We shall here go back a few years to trace the development of this set of problems and how Nambu had found himself in relationship with them.

4.2.1 Cultivating Parallel Problems: Nambu's Stereoscopic View

As the two-year-long residence in Princeton was approaching its completion in mid-1954, Nambu was prepared to return to Osaka. He was still officially affiliated with that university and he had promised to go back to upon his departure (Ashrafi 2004). While that clearly meant "to go back to the same miserable living conditions" and relative intellectual isolation of Japan after a taste of the "heavenly" living conditions in America which he and his family had meanwhile experienced,[28] he was about to do so given his search for an academic position had not had a positive outcome. In fact, another force was pushing him to stay, the force of pride. He felt that he should have tried to get another opportunity to do "good work" in a place different from Princeton before leaving. To his eyes, the stay at the Institute for Advanced Studies had been unproductive and had frustrated almost all of the expectations he had arrived with (Ashrafi 2004). Beside the adjustment to the new living conditions, he was blaming the outcome on his increasing discomfort with the highly competitive and stifling research environment he had found there. If, on the one hand, he knew that during the stay he had constructed a powerful multi-purpose formalism (his collective description of many-particle systems), on the other hand, he had not managed to solve with it the problems of the nuclear forces and the properties of nuclei which he had set up as the main goal, nor he had found another suitable direct application for it (see Sect. 3.2.2).

It is in this mix of dissatisfaction and hurt pride that a job offer finally comes, rather unexpectedly and on short notice, to positively upset Nambu's plans (Ashrafi 2004). The theoretical physicist Marvin Goldberger from the University of Chicago—who had been a doctoral student of Enrico Fermi at the same university—proposes Nambu as a post-doctoral fellow at the physics faculty of the same university and the related Enrico Fermi Institute for Nuclear Studies. The environment Nambu finds at the

---

[28] Curiously, Nambu's description is similar to the one that Tomonaga had previously given in a letter to his peers to describe his state of mind at the Institute for Advanced Studies: "isolated in paradise". Again, when talking about the campus of Chicago University on another occasion, Nambu says "its atmosphere was like a paradise" (Nambu 2016). These spontaneous references are quite eloquent when taken as expressing, by way of contrast, the life they had lived in Japan.

Institute there is, with regard to the atmosphere, antipodal to the one he had just left in Princeton: relaxed, familiar, collaborative, and marked by the mixture of disciplines and expertise—ranging from solid-state physics to nuclear and particle physics—more affine, aside from the material resources, to that Nambu had left back in Japan. The range of disciplines, but most of all the fostering of frequent interactions between them, had been a characteristic feature of the Institute since its foundation in the immediate post-war years, and was promoted by the General Seminar that Enrico Fermi had established as the first director and co-founder (Bonolis and Melchionni 2003; Freund et al. 2008; Nambu 2016).

The group Nambu joins is directed by the senior scientist Gregor Wentzel. Wentzel is a German mathematical physicist with wide interests and, coming from the school of Arnold Sommerfeld, at that time mainly occupied with strong coupling models of nuclei and superconductivity (Freund et al. 2008). "To forget the two years of bad dream", Nambu begins his research activity in Chicago by embracing dispersion theory in various occasions collaborating with Francis Low and Geoffrey Chew at the neighbouring University of Illinois (Nambu 2016).[29] At the time, these particle physicists are part of a circle in which dispersion theory is believed to be the most promising technique. In a later recollection, Nambu would call the University of Chicago of the period as "a mecca of dispersion theory" (Nambu 2016), most probably alluding to both the positive and non-edifying acceptations of the expression. Dispersion theory does, in effect, present the traits of a scientific hype in that moment. A passage from a speech given in 1964 by Leon van Hove—at the time leader of the Theory Division at CERN—conveys some of it (Van Hove 1964):

> Until some really new and fundamental discoveries are made and truly confirmed, one should avoid all claims that some formalism is the key to the situation and is superior to all others. Such claims are fully unjustified. In the very complex situation we are facing both in elementary particle physics and in nuclear physics, the large variety of formalisms and the possibility of shifting from one to the other offers great possibilities of complementary approximation schemes and should be exploited in full in the future, as had been done in the past [...]. It has actually been the case in elementary particle theory that the use of dispersion relations has been put forward at one time with very great force as being superior to other techniques, and some physicists went to very great efforts to convince their colleagues that everything can be done with dispersion theory [...]. The achievements of dispersion theory are undeniable in giving a rough correlation between many data. Quantitative results are, however, very few and the help of other formalisms has been and remains essential.

These considerations seem to reflect partly the position held by Nambu towards these techniques (Nambu 1995a). Coherent with the way of doing physics that had hitherto characterised his research activity, and at variance with the attitude of other dispersion theorists, Nambu's interest in them does not seem to stem from such a full and unconditioned adhesion. Dispersion theory was *one* of the mathematical methods that Nambu happened to collect and cultivate, which helped to rationalise *one* aspect of physical reality. Led by his inclinations, Nambu in effect ends up exploring the mathematical structure of the theory, more than he is motivated by the instrumental

---

[29] Dispersion theory is a formalism to model scattering amplitudes in particle processes which had been introduced by Goldberger himself and Murray Gell-Mann.

use of the techniques in the analysis of the data (Ashrafi 2004). Indeed, although active in the realm of dispersion theory, Nambu's pluralistic attitude towards formal tools and his "sensitivity" to problems in other contexts will not take long to concretely manifest itself again, triggered by an "accidental" event, as Nambu would later call it (Nambu 1989). The opportunity presents itself when, in early 1957, Wentzel invites Robert Schrieffer, from Bardeen's group at the neighbouring University of Urbana-Illinois, to give a seminar in Chicago on the latest results in the search for the microscopic mechanism of superconductivity. The seminar opens the branch that we have already reconstructed and which will conduct Nambu in mid-1959 to unveil the logical connection existing between energy gap, gauge invariance, and collective excitations in superconductors (see Sect. 4.1.5). This will unexpectedly turn out to be a pivotal picklock for the other problem he had at hand in particle physics. At the time of the seminar, however, superconductivity is for Nambu still a parallel and certainly lateral branch to his current main research interest. Officially employed as particle physicist, in Chicago his research activity is on elementary particle theory, and the problems that were at the moment standing in the way of a unified view of particles and interactions. One such problem, that had taken the centre of the stage right in early 1957, is triggered by the quite sensational results of an experiment that the Chinese-American physicist Chien-Shiung Wu and her team had carried out in December of the previous year (Wu 2008).

The Violation of Mirror Symmetry and the Search for a New Symmetry of Physics

One of the rare women physicists at the time even in her country of adoption, fine experimenter and a "worldly, elegant, and witty" person (Jiang et al. 2014), Chien-Shiung Wu had set up an experiment to measure the decay/transmutation of unstable nuclei of cobalt into stable nickel nuclei, electrons, anti-neutrinos, and high-energy photons. The experiment evidenced that most of the electrons out of the decay were emitted along a very specific direction in space relative to the nuclear spins, rather than being emitted evenly in all directions.[30] That implied that the laws governing weak interactions were not invariant under parity inversion (i.e., the simple change of the sign in the spatial coordinates from $\vec{x}$ to $-\vec{x}$), and thus that parity was no longer a conserved quantity in weak interaction processes. Physically speaking, differently than all the other interactions hitherto known, weak interactions distinguished between the right-handed and left-handed "version" of a particle—that is, a particle and its

---

[30] It is worth remarking here that the experiment would not have been possible without cooling the sample of cobalt salt to liquid helium temperatures—and thus the same technology that had enabled Kamerlingh Onnes to discover superconductivity 45 years earlier—as the reliability of the outcome rested on the alignment of the nuclear spins (Wu 2008). The moral is that in these two cases, among many others, it is only by eliminating a large part of the thermal motions (temperature fluctuations) that one enables the subtler effects, hitherto invisible, to become visible.

mirror image. This for instance implied either that neutrinos, massless particles subject to weak forces, exist only in the left-handed form (i.e., only in one of the two possible chiral states), or that neutrinos in the right-handed form do not take part in any interaction. This was what the outcome of Wu's experiment showed beyond any reasonable doubt.[31]

Conserved quantities—or "charges" in analogy to electric charge—and the corresponding symmetries single out what remains constant in what transforms, and serve thus as a "pin" to constrain and model transformation (at least ever since the invention of the formalism of Lagrange and Hamilton). This way of formalising physical processes is especially relevant in particle processes—such as the interaction between two particles, or the transmutation of one into another as was the case for Wu's experiment. This is because the force (in the Newtonian sense) bringing about the transformation cannot be directly observed, and thus the only way is to identify and use what remains constant in the process.

If parity was no longer a conserved quantity in weak particle processes, for the theoreticians who believed in the correctness of the experiment the tasks were quite clear. The first would be that of identifying a new candidate for the conserved quantity/symmetry. Using that symmetry as an element, the second task would be that of producing a new theoretical framework for weak interactions which would include the observed parity breaking. Weak interactions had hitherto still been based on a phenomenological model proposed by Fermi in 1933, and now the parity violation could be used as a theoretical lever to restrict the range of possible weak couplings that were compatible with Fermi's model. The first proposals along these lines were put forward by Tsung-Dao Lee & Chen-Ning Yang (1957) and Adbus Salam (1957) as soon as Wu had obtained her as yet preliminary and still unpublished results. In effect, it was Lee & Yang themselves who had raised in early 1956 the question of the invariance of weak interactions under space inversion (P, switching left and right), charge conjugation (C, switching particle and antiparticle), and time reversal (T, changing the direction of time). To this suggestion Wu had reacted by devising the experimental setting that would enable testing for the first of those symmetries (Lee and Yang 1956). Both proposals showed that if parity (P) was not conserved, charge conjugation (C) too was automatically not, but the product of the two (CP) could have been conserved in weak processes. This product, defining a quantity called chirality, would have been the natural candidate as the new conserved symmetry/quantity in weak interactions. Lee & Yang, and Salam's models proved that to be indeed the case for neutrinos, provided they were massless. On the other hand, massive particles like nucleons were also subject to weak interactions, and by the same token, as Salam mentioned in his proposal, those would automatically violate chiral symmetry. This brought to the fore for the first time a conflict between chirality and mass, conflict that will prompt Nambu to draw the analogy with superconductivity, as we will see.

---

[31] As far as the reconstruction in Hudson and Lide (Hudson and Lide 2002) goes, and as commonly happens for experimental claims of such importance, the confirmation that dispelled the residual doubts about the result came when other groups managed to obtained the same results in the course of 1957.

On the front of the modelling of weak interaction processes, the problem was to determine the combination of couplings that would be consistent with the plethora of experimental data on decay processes and compatible with the parity breaking. This had to wait until 1958, with the models proposed independently by the two duos of Robert Marshak & George Sudarshan (1958), and Richard Feynman & Murray Gell-Mann (1958). Those models, known as V − A models, hypothesised that weak processes took place via the direct interaction of nucleonic and leptonic currents, and that such currents had a vector component (V) and an axial-vector, or chiral, component (A).[32] Consistent with the available experimental data, the models convincingly suggested the vector current to be conserved in weak processes, and gave to the conservation a physical interpretation by way of an analogy with quantum electrodynamics (Feynman and Gell-Mann 1958). On the other hand, as far as the chiral current went, its conservation remained only a desirable hypothesis and the interpretation of it yet unclear. Extending the conservation to the chirality, in effect, run into the aforementioned incompatibility between such conservation and mass already raised by Salam. The chiral current of a (massive) nucleon as such did not conserve chirality in the first place. To be sure, one could construct a current, J, that conserved chirality (Polkinghorne 1958; Goldberger and Treiman 1958a). This was obtained by adding to the chiral current of a nucleon, J_N, an additional term obtained on purely *formal* grounds on the requirement that the divergence of the modified current, J, would be zero. The additional term featured the mass of the nucleon, m, and a massless pseudoscalar field, π. In mathematical terms:

$$J_\mu = J_N - 2m\partial_\mu\pi \quad \Rightarrow \quad \partial_\mu J_\mu = 0.$$

If this current formally conserved chirality, however, the *physical* reason and meaning of the additional term were unclear nor really understood.

Another element, somewhat cryptic, was added to the picture by Marvin Goldberger—Nambu's collaborator in Chicago—and Sam Treiman, at the end of 1958 (Goldberger and Treiman 1958b). Assuming the formally-constructed current above and performing an analysis by means of dispersion relations of the decays of the pion and the nucleon, the two derived an enticingly simple formula relating the characteristic constant of the *weak* forces (the pion lifetime and the axial-vector coupling constant) with those of the *strong* force (the pion-nucleon coupling constant). That was indeed an intriguing prediction, but due to the way it had been derived, it was to many physicists a rather mysterious and isolated fact, and as such it stimulated the search for a sounder derivation (Treiman 1996).

Such were the developments that had been set in motion by the breakthrough of parity violation and the necessity of identifying a new suitable symmetry of weak and strong interactions. For most of those who worked on these issues, it would have

---

[32] The vector and axial-vector components contribute equally and with opposite sign, wherefrom the shorthand V − A. Axial-vector and chiral are used as synonyms.

required a stretch of imagination to conceive that a way out of the impasse would come from the outside, as the *deus ex machina* of Ancient Greek theatre falls from above.

4.2.2 The Vacuum of the Universe Through the Lens of Superconductivity

The comment that Nambu made in the summer of 1959 at the Rochester Conference (see Sect. 4.2) refers to a way to solve in one shot the theoretical issues that followed the discovery of parity violation. These were (1) the incompatibility between the conservation of chirality and the mass of the fermions that participated in weak interaction processes; (2) the physical meaning of the additional current in the expression for the nucleon chiral current; and (3) the shaky derivation of the Goldberger-Treiman relation. What makes Nambu think that he has the solution, even while having just a sketch, is the striking formal analogy he had recognised between the expression for the *electric* current of the dressed *quasiparticle* in a superconductor, J^{sc}, and the expression for the *chiral* current of a *nucleon*, J (Nambu 2010). These were respectively given by the two formulae (which we have already reported separately above, in their respective contexts):

$$J^{sc} = J_{QP} - 2\phi\, e\, \partial_\mu f \quad \text{and} \quad J = J_N - 2m\partial_\mu \pi. \tag{4.2}$$

The first expression figured in the mathematical reformulation of superconductors which Nambu had just completed, and submitted (Nambu 1960e). There, the additional term $-2\phi\, e\, \partial_\mu f$ represented the electrical current associated with the collective longitudinal mode, f, and descended as a *necessary* consequence of the emergence of the energy gap (ϕ) from a gauge-invariant/charge-conserving dynamical equation (see Sect. 4.1.5). As far as the expression for the chirality-conserving nuclear current, Nambu was aware of its existence as a close collaborator of Goldberger and having himself worked on dispersion relations.

Through this stereoscopic view on the expression of the nuclear current granted by the knowledge of the parallel issue in superconductivity, Nambu recognises a set of analogical correspondences practically at a glance: the correspondence between the nucleon mass (m) and the energy gap, the massive nucleon and the Bogolyubov quasiparticle, and that between the two additional currents ensuring the conservation of charge and chirality, respectively. The general consequence of the identification between the structures of the two distinct domains follows suit: if the mass of the nucleon, like the energy gap, is entirely generated from a basic massless fermion field and an underlying interaction conserving chirality, then collective modes of the form appearing in the expression for the nuclear current necessarily arise. This immediately endowed the formula for the chirality-conserving nuclear current with a certain physical cogency (Nambu 1960a).

The set of correspondences suggested by the analogy between the conserved currents matched those that Nambu had noticed a few months earlier between the energy spectrum of Bogolyubov quasiparticles and that of fermions with mass m,[33] along with the corresponding equations describing them[34]:

$$E = \pm\sqrt{\epsilon_p^2 + \phi^2} \qquad E = \pm\sqrt{p^2 + m^2},$$

$$H_0'\psi_{p,+} = \epsilon_p\psi_{p,+} + \phi\psi_{-p,-}^{\dagger} \qquad H_0'\psi_R = \sigma \cdot p\psi_R + m\psi_L$$

$$H_0'\psi_{-p,-}^{\dagger} = -\epsilon_p\psi_{-p,-}^{\dagger} + \phi\psi_{p,+} \qquad H_0'\psi_L = -\sigma \cdot p\psi_L + m\psi_R$$

In light of the interpretation of the ground and excited states that Nambu had adopted (from Dirac) to formalise the solids (see Sect. 3.2.1), all these formal analogies cease to appear just as a curious coincidence, and start acquiring a consistent physical meaning. If in the superconductor a quasiparticle can be excited out of its ground state by providing an energy $E \geq \phi$, just so a massive particle is excited out of the vacuum by supplying an energy of $E \geq m$. As the second and third pairs of equations convey, if quasiparticles in the superconductor are a superposition of (gapless) electron ψ_e and hole ψ_h emerging when $\phi \neq 0$, then massive fermions in the universe arise as a superposition of massless fermions with left-handed ψ_L and right-handed chirality ψ_R and their mass is entirely dynamical.

Leveraging these formal correspondences, a hitherto unseen hypothetical "structure" of the vacuum of the universe and its excitations (the elementary particles) can be inductively inferred, unfolds and becomes visible. As the quasiparticles are elementary excitations of the ground state of the superconductor which violates gauge symmetry, the massive fermions would then be the elementary excitations of a vacuum of the universe which violates chiral symmetry. As the former spectrum is generated from "simple" electrons and a gauge-invariant attractive phonon interaction between them, the latter would be generated from simple massless fermions and some chiral-invariant attractive primary interaction. Following on with tracing the analogical correspondences, a counterpart of the massless collective modes that necessarily emerged as a consequence of the energy gap in the superconductor should have emerged as a consequence of the particles' mass, too. In the superconductor, these were the quanta formed by pairs of quasiparticles bound by the same attractive phonon interaction that was responsible for the quasiparticles. When considering the nucleon for the massive fermion, as in Eq. (4.2), the analogical counterpart of these

---

[33] This analogy between the energy spectra first appears in a note from the beginning of 1959 (see Sect. 4.1.5), and then in Nambu's paper on superconductivity (see the remarks between Eqs. 2.14 and 2.16 in Nambu 1960e).

[34] The Hamiltonian operator in the second and third line has the form $H_0' \equiv H_0 + \Sigma$, where the H_0 is the kinetic energy and Σ is the self-consistent self-energy. In the case of the Bogolyubov quasiparticle (left column), Σ is the operator defined in Eq. (4.1), featuring the self-consistent Hartree potential ϕ which represents the energy gap. In the case of the massive fermion (right column), Σ features an analogous Hartree potential m representing the mass. Like in the superconductor, such potential is determined by linearising a non-linear four-fermion interaction (see Sect. 4.2.3).

pairs would be pairs formed by a nucleon and an antinucleon bounded by the same hypothetical attractive primary interaction. In line with the composite model that was proposed by Fermi and Yang in 1949, such pairs would correspond to the pions that mediated the nuclear forces,[35] and appear in Nambu's picture as the pseudo-scalar analogues of the longitudinal collective modes, essential to the chiral invariance of weak processes and arising as a necessary consequence of the same interaction that is responsible for the nucleon mass. As Nambu understands—and shows a few months later (Nambu 1960a)—looking at the pion in this way would have also allowed one to arrive at the rather mysterious Goldberger-Treiman relation, now from a route different than the one taken by the two proponents and underpinned by a clear physical interpretation.

This is how the superconductor, as a sort of microcosm, contained a mechanism that was possibly at work in the universe at large and could thus be used to infer a new relation existing between a set of its elements. This is the sense in which the universe is *like* a superconducting medium, and the analogy makes visible a hypothetical but compelling relation between the former's physical manifestations which solves an apparent incompatibility between them. As much as an analogy can be compelling, however, it is at most just plausible. By way of an analogy, we can infer the structure of something by means of something else in which that structure happens to be known or visible, but not prove the truth of that inference. In fact, as we lay out elsewhere (Furlan and Gaudenzi 2022), the primary function and value of analogical reasoning lie in the generation of a hypothesis, the burden and honour of the proof of such hypothesis being left to other tools of rationality. As such, the "structure of the universe" that had unfolded in Nambu's head was still "just" an hypothesis. To give it some concreteness, potential testability, and more analytic as well as rhetorical persuasiveness, one needed to show that some sensible "superconducting model" of elementary particles could actually be constructed. Obtaining some direct or indirect testable predictions out of such a model would then be the proof that the analogy had led to a correct hypothesis.

In light of what Nambu already pictured, it was quite natural for him to proceed with the construction of a dynamical model of nucleons and mesons. It is through such analogical application that the symmetry breaking peculiar to superconductivity detaches itself from its original context of use to enter the field of particle physics and, at the same time, to be recognised as a general concept.

4.2.3 From Flash to Flesh: The Theory of Elementary Particles Suggested by Superconductivity

The mathematical and interpretive implications that had flashed before Nambu's eyes in mid-1959 upon identifying the vacuum with a superconductor were indeed

---

[35] Here we are simplifying for the sake of the exposition. In fact, these would be *massless* analogues of the pions (see Nambu 1960a).

enticing. At the most general level, they sketched out a blueprint for a unified under-standing of the baryons from some basic field. Such an aim at unification was, in effect, in line with the attempts made by particle physics in the last decade and which had been in the air for at least as long. As we have seen (Sect. 2.1.4), one can consider the discovery of many new particles in the immediate post-war period and the composite model of Fermi and Yang as starting points for those attempts. Among others, Werner Heisenberg had seriously undertaken the task from the early 1950s, and in 1958 he presented a *Weltformel*—literally, a "world formula"—which purported to derive all the elementary particles from just one single fermion field (the so called "ur-fermion") interacting with itself.[36] In the context of the build-up of "tension" towards a "new theory the birth of which is expected by all of us so eagerly" (Tamm 1960)—as Igor Tamm had expressed in the concluding remarks of the afore-mentioned 1959 International Conference for High-energy physics in Kiev—which characterised the grand particle physics conferences of the time, Nambu believes that his idea might offer a key. More modestly and realistically than ambitious schemes like Heisenberg's, Nambu's aims at illustrating the potential of his analogy-driven idea by "testing" it on a simplified model that tried to account for the nucleons and mesons in a unified way.

With this objective in mind Nambu starts working in collaboration with Giovanni Jona-Lasinio. A recently-graduated 27-year-old Italian mathematical physicist from Rome, Jona-Lasinio had joined the Fermi Institute in Chicago in September 1959, eager to work with Nambu (Jona-Lasinio 2016). The two had met in Rome a few months earlier, where Nambu—on his way to Kiev—had given a seminar. Jona-Lasinio's collaboration comes right when Nambu has the plan but not enough mental resources to develop it due to the grave illness of his son (Ashrafi 2004). For his part the young Italian physicist later recalled having found in Nambu for the first time a way of thinking that was congenial to him and would form his intellectual taste and inclinations ever after (Bonolis and Melchionni 2003). "Nambu was sensible to analogies", Jona-Lasinio would recollect many years later, "and I discovered in that occasion that the juxtaposition of ideas coming from distinct disciplines [in the specific case, solid-state physics and particle physics] was to me natural, and actually a possibility that fascinated me very much" (Bonolis and Melchionni 2003).[37]

What Nambu and Jona-Lasinio end up building is quite literally "a 'superconduc-tor' model of elementary particles". As evidenced by a seminar with this same title

---

[36] Heisenberg presented his theory in various occasions throughout the decade 1950–60. For an accessible summary see, e.g., Heisenberg (1957). Alexander Blum (2019) has published a detailed and critical reconstruction of the way to Heisenberg's *Weltformel*, the position and role of Pauli in it, and the debate about it in the scientific community. It is worth remarking that Nambu himself had been seduced by ideas similar to Heisenberg's in the early '50s when he had attempted to find a regularity in the masses of the particle and reviewed various hypothetical models of elementary particles (Nambu 1952a, b).

[37] The passage is translated from Italian. As Jona-Lasinio later remarked on various occasions, most of his subsequent significant works were informed by analogies, cross-fertilisation and transfers of ideas between branches of physics (see, e.g., Jona-Lasinio (2003). As he says in the aforementioned recollection, he "had radicalised Nambu's lesson".

held half a year later at Purdue University,[38] the two adopt one-to-one the many-body field theoretical framework and approximation procedure that Nambu had just applied to superconductors, and earlier to plasmas and nuclei (Nambu 1995b). As for the dynamical equation, Nambu and Jona-Lasinio settle for the one employed by Heisenberg in his non-linear spinor theory of matter, where now the spinor self-interactions play the role of the four-fermion interparticle interactions in many-body systems and Heisenberg's ur-fermions are re-interpreted as simple bare nucleons. Taking the expectation value of Heisenberg's non-linear interaction term following the procedure earlier used for plasma and superconductors and equating this expectation value to the nucleon mass,[39] Nambu and Jona-Lasinio obtain a self-consistent equation for the latter that has a non-trivial solution when the coupling constant is negative (that is, the non-linear interactions are assumed attractive). This shows that Heisenberg's equation indeed features a "superconducting solution": a mass $m \neq 0$ for the nucleons is generated entirely as a consequence of the interactions and has the same *formal* characteristics as the energy gap solution—among others, that characteristic non-analyticity that would make it impossible to derive with a perturbative approximation. To this non-trivial solution, a certain vacuum state of the world Ω_m corresponds whose energy lies below the vacuum Ω_0 associated to the trivial solution $m = 0$ (representing a world in which the nucleons do not have mass). In that it has a higher energy than the analogue of the superconducting ground state Ω_m, the vacuum state Ω_0 is, Nambu and Jona-Lasinio write, "unstable like the normal state below the critical temperature for a superconductor". As the two physicists show, there in fact exist as many such equivalent vacua $\Omega_m^{(\alpha)}$ with the same energy as there are chiral-dependent possible solutions for the mass, as occurred with the gauge-dependent gap in the superconductor.[40] In virtue of this dependence, these vacua and the quasiparticles that arise as its excitations (massive nucleons, in this case) violate the chiral symmetry of the dynamical equation they descend from. As a necessary consequence of that and in analogy with the superconductor, pairs of quasiparticles emerged that were tentatively identified with the mesons.

All these results were closely based on Nambu's reformulation of superconductivity. Yet, by way of this application to elementary particles, Nambu and Jona-Lasinio explicate some aspects that had remained implicit in that reformulation, and in this way work towards the generalisation of the mechanism originally evidenced in super-

---

[38] The seminar was held in April 1960. Nambu was unable to present this material in person due to the health issues of his son, and Jona-Lasinio presents it in his stead (Ashrafi 2004).

[39] Heisenberg's non-linear interaction term is equal to $-g[\bar{\psi}\psi\bar{\psi}\psi - \bar{\psi}\gamma_5\psi\bar{\psi}\gamma_5\psi]$. In analogy to the procedure already adopted by Nambu in his collective description (see Sect. 3.2.2), the self-consistent equation for the mass is obtained by linearising the interaction and defining the self-consistent mass coefficient according to $-2g[\langle\bar{\psi}\psi\rangle\bar{\psi}\psi - \langle\bar{\psi}\gamma_5\psi\rangle\gamma_5\bar{\psi}\psi] := -m\bar{\psi}\psi$. The resulting equation for the mass $m = 2g[\langle\bar{\psi}\psi\rangle - \langle\bar{\psi}\gamma_5\psi\rangle\gamma_5]$ is self-consistent since $\langle\bar{\psi}\psi\rangle = S_F^{(m)}$ is the propagator for the fermion with the mass m and thus itself depends on the sought-for mass (for further details see Eqs. 3.4–3.6 in Nambu and Jona-Lasinio (1961)).

[40] The quantity $\alpha \in [0, 2\pi]$ is the phase of the chiral transformation of the fermion field $\psi \rightarrow e^{i\alpha\gamma_5}\psi$. Each specific value of α identifies one specific vacuum state solution, or "world state", and the chiral-dependent value that the mass has in that state $me^{2i\alpha} = m(\cos(2\alpha) + i\gamma_5 \sin(2\alpha))$.

conductors. One such aspect is the presence of infinite ground (or vacuum) states each with its related excitation spectrum; and the fact that these are physically distinct and outside of/orthogonal to each other. Each of them is a separate "world" in that no standard (i.e., local) physical operation enables the transition to any of the others. One must therefore conclude that our universe is *one* of them, with a fixed and given value of α. This is the analogue of the superconductor having a fixed value of the gauge phase, φ, in its ground state or, as Nambu and Jona-Lasinio suggest in a more concrete example, a magnet being magnetised along *one* of the possible directions. The collective excitations, created by locally perturbing the system (e.g., rotating a finite number of spins of the whole, ideally infinitely-extended, magnet), represent then the various ways in which the system can be made to *locally* move about that particular ground state/preferred direction. In this sense, the collective excitations are the remnants of the original symmetry, the way in which that hidden symmetry manifests.[41]

In disclosing these general aspects of the symmetry breaking mechanism discovered in superconductors, this first transfer to elementary particles has two pivotal implications. These are packed into the concluding remarks of the talk held at Purdue (Nambu 1995b):

> We have to admit that we do not yet know what type of Lagrangian can possibly give rise to the observed baryon and meson spectrum. How much internal degree of freedom do we need to start with in order to account for the observed particles (and not to account for non-existing ones)? [...] Can we assume a high degree of symmetry properties in the beginning and yet come out with a smaller amount of apparent symmetries? The last question seems particularly relevant since we have found an example in the conflict between γ_5 [chirality] invariance and finite mass. Here it should be helpful to seek analogies in solid state physics for better physical understanding.
>
> For example, it is not hard to foresee how nature can manifest itself in an unsymmetric [sic] way while keeping the basic laws symmetric if we compare the situation with ferromagnetism. In an ordinary material, the ground state of a macroscopic body has spin zero practically, so that there is no preferred axis in space. In the ferromagnetic case, on the other hand, all the spins are parallel in the ground state, and they must point in some direction, thereby creating an asymmetry in reality. Such spontaneous polarizations may be happening in the world of elementary particles too.

One implication consists in the recognition that the mechanism at work in superconductors is one instance of the general mechanism of "spontaneous polarisation", as they call it, that characterises phase transitions at their core. The transition considered there is the one undergone by an ordinary material from the non-magnetic to a magnetic phase where the atomic spins all point in one among the possible direc-

---

[41] The "local" expresses here the important fact that any physical operation that one can possibly perform on the system, and by means of which one can excite it from its symmetry-broken ground state, affects always just a finite region of space, however large. The operation becomes "global" only in the limit for the extension of this region going to infinity, in which case the system could transition to other ground states (in the case of a magnet, this is represented by the coordinated rotation of all the spins). While such an infinite-wavelength collective excitation costs no energy, it cannot be physically performed, wherefrom the impossibility to transition to another ground state.

tions.[42] That one direction among the infinitely degenerate and equally possible ones represents a "world" or "reality" in which the material has acquired a certain property (a magnetisation with a given intensity and direction). This entirely dynamical property and the associated state are a non-trivial solution of the equation describing the system of interacting spins, and break the equation's original (rotational) symmetry. This happens in the same peculiar way in which it happens in the transition to the superconducting phase, with the advantage that in a ferromagnet the phenomenon is more immediately visualisable.

The second implication is that this same mechanism is suggested to be at work in the universe at large, and in a way that the properties of the particles that we see—the mass of the nucleons and the presence of mesons in this particular case—are similarly dynamical and the result of spontaneous symmetry breaking. This offers a new and quite original way of looking at the elementary particles, and enacts that crucial shift of attitude that we alluded to at the end of the section on superconductivity (Sect. 4.1.5). The shift from using the mechanism discovered in superconductivity to solve a *problem* of incompatibility between manifestations (gauge invariance and energy gap, chirality invariance and mass) to its use as a theoretical *resource*. More specifically, this is the use within particle physics of that "spontaneous polarisation" as the paradigmatic mechanism for the dynamical generation of complex and inhomogeneous manifestations and properties out of simpler and homogeneous ingredients. This new possibility for particle theory "to be richer and more complex than hitherto envisaged" (Nambu and Jona-Lasinio 1961), as Nambu and Jona-Lasinio would write, became visible only when illuminated from the outside, that is, from the realm of solid matter and phase transitions. In turn, in a sort of feedback loop, it is while working on this application to particle theory that Nambu starts realising the generality of this mechanism in solid-state physics. As it is testified by one of his notebooks, he begins to tentatively collect analogue instances of "hidden and broken symmetries" in systems other than superconductors and ferromagnets (Nambu 1962). In this program, which will though never be pursued much further, as such instances figure "superfluids", "nuclei", and "crystals".

These conceptual steps, the end of our reconstruction, mark the point where spontaneous symmetry breaking comes to light as a general concept and, even if not yet by the exact name we call it today, emerges to the surface of history.

4.2.4   Epilogue. The Impact of Nambu's Vision and the Offshoots of the Analogy

As can be deduced from the reported discussion session of the informal presentation given by Jona-Lasinio in Purdue, the very first reaction to these ideas is one of mild interest. Significant and palpable is instead their impact at the 1960 International Conference of High-Energy Physics; a "big splash", according to Nambu's later

---

[42] As is well-known the spontaneous magnetisation of a material occurs below the Curie temperature.

recollection (Ashrafi 2004). Nambu figures in the session "Theories of elementary particles" beside two other speakers and right after Werner Heisenberg. Nambu's talk is quite ambitiously entitled "Dynamical theory of elementary particles suggested by superconductivity" (Nambu 1960c). It is worthwhile to report here Heisenberg's reaction to it as it appears in the *The New Yorker* magazine (Shortell and McKelway 1960). Pressed by the journalist's questions, Heisenberg says:

> "Non linear spinor theory is an attempt to formulate a law that will explain and predict the masses and characteristics of all the elementary particles. [...] The trouble is that we have split it [the problem] into too many complex details. People are confused today. They have lost the sense of harmony that underlies everything. They do not see the big connections." According to Prof. Heisenberg, the big connections in the world of high-energy physics are to be deduced not from the properties of the elementary particles but from the relationship between them. [...] "The particles in themselves are not fundamental. [...] For myself, I must start not from detail but from a general connection, a feeling I have about the way things should be." [...]
> We asked professor Heisenberg if the Rochester Conference had provided any developments that would substantiate his belief in the equation [Heisenberg's non-linear spinor equation]. [...] "I will tell you a story that illustrates how we physicists can sometimes make progress. At this conference, Nambu read an interesting paper concerning a theory of elementary particles suggested by superconductivity—a baffling phenomenon of solid-state physics, which is entirely different than my own. [...] In working out some of the problems posed by this phenomenon, Nambu has encountered mathematical difficulties similar to those confronting me in the non-linear spinor theory. Perhaps these puzzles that seem so different are related. Perhaps, through cross-research, we can achieve some enrichment of our mathematical tools that will enable us to go on more easily. Our Russian colleague Bogolyubov is interested in the apparent link between Nambu and me. So now we must try very hard to find this link. The mathematics will be enormously complicated, but this is an unusual challenge." [..] "I am convinced that the universe is connected by a truly simple law."

If nothing really concrete follows through from the pursuit of a by-then perhaps too stubborn and ambitious Heisenberg, two other young physicists, Jeffrey Goldstone and Philip Anderson, will reap the fruits of Nambu's vision. The two distil from it different aspects and bridge it to its wider applications in solid-state physics and particle physics, among which is the Higgs mechanism. At that time a visiting fellow at CERN, Goldstone reads the ideas we have exposed on a report for the Enrico Fermi Institute in Chicago that Nambu had filed in correspondence to the talk given at Purdue (the report is cited in the first line of Goldstone (1961)). These stimulate a practically immediate reaction in him, which he articulates in the now famous paper entitled "Field theories with 'superconductor' solutions" and submitted to *Il Nuovo Cimento* in September 1960 (Goldstone 1961). For various field-theory toy models, all featuring fermions interacting via a boson field, Goldstone investigates the conditions of existence of superconducting solutions. These are indeed shown to have symmetries lower than the model equation they derive from and to be accompanied by massless quanta. Generalising these results, in early 1962, Goldstone, along with Steven Weinberg and Abdus Salam, formulates the mechanism discovered by Nambu into a compact and general theorem (Goldstone et al. 1962). The thesis is contained directly in the few lines of the abstract of their article, quite enticingly entitled "Broken symmetries":

[…] if there is continuous symmetry transformation under which the Lagrangian is invariant, then either the vacuum state is also invariant under the transformation, or there must exist spinless particles of zero mass.

The spinless particles of zero mass referred to by the authors are the quanta identified by Nambu. For a long time simply called "Goldstone bosons", they will only later acquire the more appropriate name of "Nambu-Goldstone bosons" that they have now (Freund 2016). This marks in name and in fact the official entrance of spontaneous symmetry breaking into the history of physics.

The mechanism that Nambu uncovered and was then given its synthetic form by the Goldstone theorem has ever since been found to be at work in a wide and wild variety of physical systems in solid-state physics, beyond the exemplary instances of superconductors and magnets. The dizzying array of names for these kinds of excitations reflect their ubiquity and significance in solid-state physics.

Anderson and the More Literal Analogy to Superconductivity

As far as particle physics was concerned, the analogy drawn by Nambu and the ensu-ing superconducting model of elementary particles would blaze the trail to analogies between solid-state systems and the vacuum of the universe and its excitation spec-trum. In itself, however, the specific analogy between gauge invariance and chiral invariance underpinning the specific model developed by Nambu and Jona-Lasinio would not have much immediate fortune. As it turned out, a more literal analogy to superconductivity than the one Nambu had been driven to would solve the problems of finding a unified way of looking at particles and interactions. The physicist Philip Anderson will actually indicate the way to it in the aftermath of the work by Nambu and Jona-Lasinio.

At the time of the events we have reconstructed, Anderson is a solid-state physicist employed at the Bell Telephone Laboratories.[43] There, he had been hitherto working on a variety of theoretical problems in the realm of solids, from studying various effects of impurities in metals to the problem of gauge invariance in superconductors (Anderson 1958a, b). Given his interests, Anderson is well aware of Nambu's refor-mulation of superconductivity and his elegant solution to that problem, as well as the ingenious analogical transposition to particle physics that Nambu had performed (Kojevnikov 1999). Nambu had spoken about these works right at the Bell Labora-tories where he had been invited for a talk by Anderson himself in the summer of 1960—and would be invited also for a summer stay in the following years (Nambu 1960d; Ashrafi 2004). Although an outsider to particle physics, as Anderson later recollected, in that period he had been quite exposed to it by way of the many par-ticle theorists—and other physicists "who were on the borderline between particle

---

[43] Bell Laboratories had various bases around the country. Bardeen was employed in Urbana, Illinois, while Anderson was employed at Murray Hill, New Jersey. We have already mentioned the role of this institution in fundamental and applied research in relation to John Bardeen's discovery of superconductivity.

and condensed-matter theory like Brout" Kojevnikov 1999)—who visited the Bell
Laboratories, as well as during the academic year 1961–62 he spent as a lecturer
at Cambridge University. Indeed between Cambridge and London, we find at the
time, for instance, the authors (Goldstone, Weinberg, and Salam) of the "Broken
Symmetries" paper mentioned in the previous section, as well as a good deal of
other physicists working on the so-called gauge theories of elementary particles and
interactions.

Gauge theories were based on the so-called gauge principle, according to which
the invariance under a given *local* gauge transformation implied the existence of a
massless vector boson that would mediate the interaction. In the case of electromagnetism, such bosons could be identified as the photons, and thereby the principle
scheme held and was experimentally adequate. Apart from that, however, that desirable scheme seemed not to be applicable to the other known particle interactions in
nature (strong and weak forces), which were instead supposedly mediated by *massive*
bosons. This was also the reason why, even assuming that these gauge symmetries
would be spontaneously broken and invoking the freshly discovered mechanism of
spontaneous symmetry breaking, would not help as any boson that could possibly come out of that mechanism would be massless, too. Two mechanisms which
hypothetically offered ways to unify elementary particles thus run into the same
inadequacy.

In this situation of impasse Anderson comes in to point out that a way to overcome
that inadequacy is once again suggested by superconductivity. As we know, Nambu
clarified that in superconductors the spontaneous breakdown of gauge symmetry
necessarily gives rise to the longitudinal collective modes. What we have not said is
that in the same work Nambu also showed that the collective modes acquire a mass
when also factoring in, on top of the phonon interaction, the long-range Coulomb
interaction between electrons. This was essentially due to the same mechanism as
in plasmas first discovered by Tomonaga and Bohm & Pines (see Sects. 2.2.1 and
2.2.2).[44] Now, Anderson realizes that this aspect, known only to those physicists who
had recently worked on superconductors, is precisely an analogue of the mechanism
that particle theorists were searching for. As he writes in a short article about it,
published in 1963 (Anderson 1963):

> The superconducting electron gas [i.e., the superconductor] has no zero-mass excitations
> whatsoever. In that case, the fermion mass is finite because of the energy gap, while the
> boson which appears as a result of the theorem of Goldstone and has zero [...] mass is
> converted into a finite-mass plasmon by interaction with the appropriate gauge field, which
> is the electromagnetic field.
> It is likely, then, considering the superconducting analogue, that the way is now open for
> a degenerate vacuum theory of the Nambu type[45] without any difficulties involving either
> zero-mass Yang-Mills gauge bosons or zero-mass Goldstone bosons. These two types of
> bosons seem capable of "cancelling each other out" and leaving finite mass bosons only.

---

[44] We remind the reader that the normal state of most superconducting materials is a plasma, that
is, a gas of electrons interacting via (long-range) Coulomb forces.

[45] Here Anderson refers to the article by Nambu and Jona-Lasinio (1961).

Used as the analogue model of a theory of elementary particles, the superconductor actually revealed a mechanism to explain not only the mass of the fermions, as found by Nambu, but at the same time also the mass of the bosons. In drawing his ingenious analogy, and lured by the prospect of the dynamical model of nucleons and mesons, Nambu had transported the former aspect of superconductivity to particle physics, but missed the latter. He had failed to see another, in fact more literal, way in which the superconductor could be used to inform elementary particles. In following the trail suggested by the formal analogy between the electric current of a quasiparticle and the chiral current of a nucleon, Nambu had analogised the gauge symmetry with chiral symmetry. In doing so, he had, however, inadvertently introduced some formal and physical disanalogies. The pions/zero-mass Goldstone bosons in Nambu's particle model were in fact no longer actual longitudinal collective modes; and the massive fermions were no longer conceived as charged quasiparticles that would, like in the superconductor, interact also via Coulomb forces. As a result of these disanalogies, the possibility of combining the collective modes arising as zero-mass Goldstone bosons in the superconductor with the gauge bosons (mediating the long-range Coulomb forces) that Anderson now sees could no longer be contemplated in Nambu's model of elementary particles.

Primed by Nambu's analogy between condensed matter systems and elementary particles, Anderson interiorises the lesson and draws a more "literal" analogy to superconductivity, one which retains and compares more integrally the physical aspects of the superconducting medium and the vacuum medium. In an analogy something is conserved, kept constant, while something else changes. A certain freedom in choosing, consciously or not, what to conserve and what to change corresponds to the fact that there are always multiple ways in which two things can be analogised to each other. It is not a question of one way being right and another wrong, but a question of what is instructive for a given purpose. Each way has both virtues and limits, leads somewhere and not somewhere else, makes one thing visible while concealing another. That is, after all, the hard life of discoverers.

References

Anderson, P. W. (1958). Coherent excited states in the theory of superconductivity: Gauge invariance and the Meissner effect. *Physical Review, 110*, 827–835.

Anderson, P. W. (1958). Random-phase approximation in the theory of superconductivity. *Physical Review, 112*(6), 1900.

Anderson, P. W. (1963). Plasmons, gauge invariance, and mass. *Physical Review, 130*(1), 439.

Ashrafi, B. (2004). Interview of Yoichiro Nambu by Babak Ashrafi on 2004 July 16. www.aip.org/history-programs/niels-bohr-library/oral-histories/30538.

Bardeen, J. (1937). Conductivity of monovalent metals. *Physical Review, 52*, 688–697.

Bardeen, J. (1951). Relation between lattice vibration and London theories of superconductivity. *Physical Review, 81*, 829–834.

Bardeen, J. (1956). *Theory of Superconductivity*, pp. 274–369. Springer Berlin Heidelberg.

Bardeen, J. (1957). Gauge invariance and the energy gap model of superconductivity. *Il Nuovo Cimento, 5*(6), 1766–1768.

Bardeen, J. (1972). Nobel lecture. NobelPrize.org. Nobel prize outreach. https://www.nobelprize. org/prizes/physics/1972/bardeen/lecture/.

Bardeen, J. (2010). Development of concepts in superconductivity. In *BCS: 50 years*, pp. 33–40. World Scientific.

Bardeen, J., Cooper, L. N., & Schrieffer, J. R. (1957). Theory of superconductivity. *Physical Review, 108*(5), 1175.

Bardeen, J., & Pines, D. (1955). Electron-phonon interaction in metals. *Physical Review, 99,* 1140– 1150.

Bardeen, J., & Schrieffer, J. (1961). Recent developments in superconductivity. In *Progress in low temperature physics* (Vol. 3, pp. 170–287). Elsevier.

Becker, R., Heller, G., & Sauter, F. (1933). Über die Stromverteilung in einer supraleitenden Kugel. *Zeitschrift für Physik, 85*(11), 772–787.

Blum, A. (2019). *Heisenberg's 1958 Weltformel and the roots of post-empirical physics.* Springer-Briefs in History of Science and Technology. Springer International Publishing.

Bogolyubov, N. (1947). On the theory of superfluidity. *Journal of Physics, 11*(1), 23.

Bogolyubov, N. (1958). On a new method in the theory of superconductivity. *Il Nuovo Cimento, 7*(6), 794–805.

Bogolyubov, N., Tolmachov, V. V., & Širkov, D. (1958). A new method in the theory of superconductivity. *Fortschritte der Physik, 6*(11–12), 605–682.

Bonolis, L., & Melchionni, M. (2003). *Fisici italiani del tempo presente: storie di vita e di pensiero,* Chapter Giovanni Jona-Lasinio (pp. 179–218). Ricerche (Marsilio editori). Marsilio.

Cooper, L. N. (2011). Remembrance of superconductivity past. In *BCS: 50 Years* (pp. 3–19). World Scientific.

Corak, W. S., Goodman, B. B., Satterthwaite, C. B., & Wexler, A. (1954). Exponential temperature dependence of the electronic specific heat of superconducting vanadium. *Physical Review, 96,* 1442–1444.

Dahl, P. F. (1984). Kamerlingh Onnes and the discovery of superconductivity: The Leyden years, 1911–1914. *Historical Studies in the Physical Sciences, 15*(1), 1–37.

Daitch, V., & Hoddeson, L. (2002). *True genius: The life and science of John Bardeen: The only winner of two Nobel Prizes in physics.* National Academies Press.

Eckert, M., Schubert, H., & Torkar, G. (1992). The roots of solid-state physics before quantum mechanics. In *Out of the crystal maze: Chapters from the history of solid state physics.* Oxford University Press.

Feynman, R. P., & Gell-Mann, M. (1958). Theory of the Fermi interaction. *Physical Review, 109,* 193–198.

Freund, P. G. (2016). Nambu at work. *Progress of Theoretical and Experimental Physics, 2016*(7), 1–7.

Freund, P. G., Goebel, C. J., Nambu, Y., & Oehme, R. (2008). Gregor Wentzel, 1–12. arXiv:0809.2102.

Fröhlich, H. (1950). Theory of the superconducting state. i. The ground state at the absolute zero of temperature. *Physical Review 79*(5), 845.

Fröhlich, H. (1954). On the theory of superconductivity: The one-dimensional case. *Proceedings of the Royal Society of London. Series A. Mathematical and Physical Sciences 223*(1154), 296–305.

Furlan, S., & Gaudenzi, R. (2022). The earth vibrates with analogies: The Dirac sea and the geology of the vacuum. *Studies in the History and Philosophy of Science (accepted).*

Gavroglu, K. (1995). *Fritz London: A scientific biography.* Cambridge University Press.

Gavroglu, K., & Goudaroulis, Y. (1984). Some methodological and historical considerations in low temperature physics: The case of superconductivity 1911–57. *Annals of Science, 41*(2), 135–149.

Ginzburg, V. L., & Landau, L. D. (2009). *On the theory of superconductivity* (pp. 113–137). Berlin, Heidelberg: Springer Berlin Heidelberg.

Goldberger, M. L., & Treiman, S. B. (1958). Conserved currents in the theory of Fermi interactions. *Physical Review, 110,* 1478–1479.

Goldberger, M. L., & Treiman, S. B. (1958). Decay of the pi meson. *Physical Review, 110*, 1178–1184.

Goldstone, J. (1961). Field theories with "Superconductor" solutions. *Il Nuovo Cimento (1955–1965) 19*(1), 154–164.

Goldstone, J., Salam, A., & Weinberg, S. (1962). Broken symmetries. *Physical Review, 127*(3), 965.

Goodstein, D., & Goodstein, J. (2000). Richard Feynman and the history of superconductivity. *Physics in Perspective, 2*(1), 30–47.

Gordon, W. (1926). Der Comptoneffekt nach der schrödingerschen Theorie. *Zeitschrift für Physik, 40*(1), 117–133.

Heisenberg, W. (1957). Quantum theory of fields and elementary particles. *Review Modern Physics, 29*, 269–278.

Hoddeson, L., Baym, G., & Eckert, M. (1992). The development of the quantum mechanical electron theory of metals. In *Out of the crystal maze: Chapters from the history of solid state physics.* Oxford University Press.

Hoddeson, L., Schubert, H., Heims, S., & Baym, G. (1992). Collective phenomena. In *Out of the crystal maze: Chapters from the history of solid state physics.* Oxford University Press.

Hudson, R., & Lide, D. (2002). Reversal of the parity conservation law in nuclear physics. In *A century of excellence in measurements, standards, and technology.* CRC Press.

Jiang, C., Chiang, T., & Wong, T. (2014). *Madame Wu Chien-Shiung: The first lady of physics research.* World Scientific.

Jona-Lasinio, G. (2003). Cross fertilization in theoretical physics: The case of condensed matter and particle physics. In *13th international congress in mathematical physics (ICMP 2000)*, pp. 143–152.

Jona-Lasinio, G. (2016). Yoichiro Nambu: Remembering an unusual physicist, a mentor, and a friend. *Progress of Theoretical and Experimental Physics, 2016*(7), 1–5.

Onnes, H. K. (1911). The resistance of pure mercury at helium temperatures. *Communications from the Laboratory of Physics at the University of Leiden* (120b), 3–5.

Onnes, H. K. (1913). Further experiments with liquid helium. h. On the electrical resistance of pure metals. VII. The potential difference necessary for the electrical current through mercury below 4.19 k. *Communications from the Laboratory of Physics at the University of Leiden* (133a), 3–26.

Onnes, H. K. (1914). The imitation of an ampere molecular current or of a permanent magnet by means of a supra-conductor. *Communications from the Laboratory of Physics at the University of Leiden* (140b), 9–18.

Klein, O. (1926). Quantentheorie und fünfdimensionale relativitätstheorie. *Zeitschrift für Physik, 37*(12), 895–906.

Kojevnikov, A. (1999). Interview of Philip W. Anderson by Alexei Kojevnikov on 1999 November 23. www.aip.org/history-programs/niels-bohr-library/oral-histories/23362-3.

Kojevnikov, A. (2008). The phenomenon of Soviet science. *Osiris, 23*(1), 115–135.

Laughlin, R. (2006). *A different universe: Reinventing physics from the bottom down.* Basic Books.

Lee, T. D., & Yang, C. N. (1956). Question of parity conservation in weak interactions. *Physical Review, 104*, 254–258.

Lee, T. D., & Yang, C. N. (1957). Parity nonconservation and a two-component theory of the neutrino. *Physical Review, 105*(5), 1671.

London, F., & London, H. (1935). The electromagnetic equations of the supraconductor. *Proceedings of the Royal Society of London. Series A-Mathematical and Physical Sciences 149*(866), 71–88.

Maxwell, E. (1950). Isotope effect in the superconductivity of mercury. *Physical Review, 78*, 477–477.

Meissner, W., & Ochsenfeld, R. (1933). Ein neuer Effekt bei Eintritt der Supraleitfähigkeit. *Naturwissenschaften, 21*(44), 787–788.

Nambu, Y. (1952). An empirical mass spectrum of elementary particles. *Progress of Theoretical Physics, 7*(5–6), 595–596.

Nambu, Y. (1952). Topics on the theory of elementary particles (in Japanese). *Butsuri, 7*(2), 77–83.

Nambu, Y. (1956). Nambu, Yoichiro. Papers, [Box 40; Folder 1], Special Collections Research Center, University of Chicago Library, Chicago, USA.

Nambu, Y. (1959). Nambu, Yoichiro. Papers, [Box 14; Folder 2], Special Collections Research Center, University of Chicago Library, Chicago, USA.

Nambu, Y. (1960a). Axial vector current conservation in weak interactions. *Physical Review Letters, 4*(7), 380.

Nambu, Y. (1960b). Discussion on Thirring's and Toushek's papers. *Proceedings, 9th international conference on high energy physics, Kiev.*

Nambu, Y. (1960c). Dynamical theory of elementary particles suggested by superconductivity. In *Proceedings of the 1960 annual international conference on high energy physics at Rochester,* pp. 858–866.

Nambu, Y. (1960d). Nambu, Yoichiro. Papers, [Box 1; Folder 4], Special Collections Research Center, University of Chicago Library, Chicago, USA.

Nambu, Y. (1960e). Quasi-particles and gauge invariance in the theory of superconductivity. *Physical Review, 117,* 648–663.

Nambu, Y. (1962). Nambu, Yoichiro. Papers, [Box 11; Folder 1-4], Special Collections Research Center, University of Chicago Library, Chicago, USA.

Nambu, Y. (1989). Gauge principle, vector-meson dominance, and spontaneous symmetry breaking. In *Pions to quarks: Particle physics in the 1950s.* Cambridge University Press.

Nambu, Y. (1995a). Research in elementary particle theory. In *Broken symmetry: Selected papers of Y. Nambu,* pp. vii–xiv. World Scientific.

Nambu, Y. (1995b). A 'superconductor' model of elementary particles and its consequences. In *Broken symmetry: Selected papers of Y. Nambu,* pp. 110–126. World Scientific.

Nambu, Y. (2010). Energy gap, mass gap, and spontaneous symmetry breaking. In *BCS: 50 years,* pp. 525–533. World Scientific.

Nambu, Y. (2016). Reminescences of the youthful years of particle physics. In *Memorial Volume For Y. Nambu,* pp. 143–159. World Scientific.

Nambu, Y., & Jona-Lasinio, G. (1961). Dynamical model of elementary particles based on an analogy with superconductivity. i. *Physical Review 122*(1), 345.

Pines, D. (2009). Biographical memoirs: John bardeen. In *Proceedings of the American philosophical society: Held at Philadelphia for promoting useful knowledge* (pp. 287–322). American Philosophical Society.

Polkinghorne, J. (1958). Renormalization of axial vector coupling. *Nuovo Cimento, 10,* 179–180.

Reif-Acherman, S. (2004). Heike Kamerlingh Onnes: Master of experimental technique and quantitative research. *Physics in Perspective, 6*(2), 197–223.

Reynolds, C. A., Serin, B., Wright, W. H., & Nesbitt, L. B. (1950). Superconductivity of isotopes of mercury. *Physical Review, 78,* 487–487.

Salam, A. (1957). On parity conservation and neutrino mass. *Il Nuovo Cimento, 5*(1), 299–301.

Sardella, I. (2012). *Storia della rottura di simmetria: dalla colonna di Eulero al Bosone di Higgs, il lungo cammino di un'idea.* Aracne editrice.

Schafroth, M. (1951). Bemerkungen zur fröhlichschen Theorie der Supraleitung. *Helvetica Physica Acta, 24*(6), 645–662.

Serin, B. (1956). *Superconductivity: Experimental part* (pp. 210–273). Berlin, Heidelberg: Springer Berlin Heidelberg.

Shortell, S., & McKelway, S. C. (1960). Tiny Cards. *The New Yorker, October 15 Issue,* 34.

Sudarshan, E. C. G., & Marshak, R. E. (1958). Chirality invariance and the universal Fermi interaction. *Physical Review, 109,* 1860–1862.

Tamm, I. (1960). Concluding remarks. *Proceedings, 9th international conference on high energy physics, Kiev,* pp. 411–416.

Treiman, S. (1996). A life in particle physics. *Annual Review of Nuclear and Particle Science, 46*(1), 1–30.

Tuyn, W., & Onnes, H. K. (1991). Further experiments with liquid helium, a. The disturbance of supra-conductivity by magnetic fields and currents. The hypothesis of Silsbee. In K. Gavroglu, & Y. Goudaroulis (Eds.), *Through measurement to knowledge: The selected papers of Heike Kamerlingh Onnes 1853–1926* (pp. 363–387). Dordrecht: Springer Netherlands.

van Delft, D. (2005). *Heike Kamerlingh Onnes: Een Biografie*. Bakker.

van Delft, D. (2008). Little cup of helium, big science. *Physics Today, 61*(3), 36.

Van Hove, L. (1964). Nuclear physics and elementary particles. *In Congres International de Physique Nucleaire, 1*, 493–500.

Warnow, J., & Williams, R. (1974). Interview of Robert Schrieffer by Joan Warnow and Robert Williams on 1974 September 26. www.aip.org/history-programs/niels-bohr-library/oral-histories/4864-1.

Wu, C. -S. (2008). The discovery of the parity violation in weak interactions and its recent developments. In *Nishina Memorial Lectures* (pp. 43–70). Springer.

Chapter 5
Historical and Epistemological Reflections

What we have narrated here walking along the "world-lines" of a few physicist-thinkers is an adventure of ideas spanning over roughly thirty years, between 1930 and 1960. This story is prominently one of virtuous interactions between disciplinary and cultural traditions or, as we alternatively put it in the introduction, between worlds of phenomena and conceptual schemes. In light of the knowledge gained, we here propose two orders of considerations. Firstly, we wish to retrace the salient stages of our story focusing on the perspective shift from elementary particles to collective quasiparticles that was introduced in high-energy physics by spontaneous symmetry breaking. In particular, we will highlight (i) how that shift, mediated by the analogy between vacuum and solid media, in turn relied on the progress that had been made in the methods of approximation of many-body systems; and (ii) how that shift rendered possible the transformation of those properties of elementary particles that were previously thought of as intrinsic into entirely dynamical and relational properties. In doing so, we will reflect on the kind of explanation that spontaneous symmetry breaking introduces in high-energy physics. Secondly, we consider more in depth a few aspects of Nambu's personal knowledge that were instrumental to the discovery of spontaneous symmetry breaking. In particular, we expand on the influences that the distinct cultural milieu of Japan had on Nambu's intellectual development and heuristics.

5.1 Individual and Collective, Micro and Macro Meet at the Confluence of Three Fronts of Physics

The period of the history of physics traversed by our story, roughly 1930–1960, features a progress in various fronts of theoretical descriptive means. On the front of fundamental theory, it stretches from an established and complete (non-relativistic)

© The Author(s), under exclusive license to Springer Nature Switzerland AG 2022
R. Gaudenzi, *Historical Roots of Spontaneous Symmetry Breaking*,
SpringerBriefs in History of Science and Technology,
https://doi.org/10.1007/978-3-030-99895-0_5

quantum mechanics to the framework of the quantum theory of fields on which the subsequent elementary particle physics, and in particular the Standard Model of Elementary Particles, would be built. On a second, seemingly purely applied front—descending as an outcome of that advancement of the theoretical framework—the three decades prominently feature a number of applications of quantum mechanics to the more complex, many-particle systems at various energy and spatial scales. The computation of the energy spectra of atoms and molecules (other than the sole hydrogen atom, which is calculable exactly), through to solid matter and atomic nuclei indeed begins in the late '20s and early'30s as a mundane, albeit utterly non-trivial, application of the already *established* quantum framework and the *known* dynamical laws[1] motivated by the need to find approximation schemes that could deal with the difficulties related to the high number of components interacting with each other and/or the strength of their interactions. In some of these systems, with the increase of the number of interacting microscopic components, new collective behaviours were found to emerge and the individual particles' properties to change. These behaviours were discovered and their effects quantified using various approximation methods. Among collective phenomena, but traditionally approached with the techniques of thermodynamics and statistical mechanics, were phase transitions, the theoretical understanding of which represented a third, separate front.

These three fronts were characterised by different degrees of fundamentality and aims as well as different spatial & energy scales. Being distinguished on this basis in the system of sub-disciplines of physics, they progressed in relative independence until they found a significant intersection/confluence point in the phenomenon of spontaneous symmetry breaking. To be sure, the independence was not complete, and localised transpositions of concepts and mathematical techniques had occurred even before spontaneous symmetry breaking. Precisely these transpositions, systematically deepened and interwoven, would be instrumental to the discovery of spontaneous symmetry breaking. In his very first presentation of the idea, held in 1960 at Purdue University, Nambu significantly comments on such transpositions (Nambu 1995):

> In recent years it has become fashionable to apply field-theoretical techniques to the many-body problems one encounters in solid state physics and nuclear physics. This is not surprising because in a quantized field theory there is always the possibility of pair creation (real or virtual), which is essentially a many-body problem. We are familiar with a number of close analogies between ideas and problems in elementary particle theory and the corresponding ones in solid state physics. For example, the Fermi sea of electrons in a metal is analogous to the Dirac sea of electrons in the vacuum, and we speak about electrons and holes in both cases. Some people must have thought of the meson field as something like the shielded Coulomb field. [...] At any rate, we should expect a close interaction of the two branches of physics in terms of concepts and mathematical techniques, which make up the content of quantum field theory. In this talk we are going to show another possibility of such an interaction, but this time in the opposite direction to what has been the general trend.

---

[1] In all of the physical systems mentioned, these are the laws of electromagnetic interaction. Of course, when it came to nuclei, the hypothetical interaction laws that these methods were supposed to approximate were phenomenological and effective in various degrees.

With a touch of self-irony and the typical understatement about his own work ("we are familiar with…", "some people must have thought…"), Nambu sketched the transfer of knowledge resources that, by means of analogical correspondences, had occurred from the front of elementary particle physics to that of "applied" many-body problems in nuclear and solid-state physics. In Chap. 3, we have reconstructed how Nambu built on and contributed to the establishment of all those "close analogies between ideas and problems", and how such analogies were instrumental to his field-theoretical approach to nuclear matter, plasmas, and then superconductors. As we have shown, these constituted a part of the whole network of exchanges that *accreted* into the final leap he was about to present: the analogical transfer of the instance of spontaneous symmetry breaking that he had discovered in superconductors to the vacuum of elementary particle physics. A transfer in the same direction had characterised Tomonaga's analogy between renormalisation and Hartree & Fock's subtraction procedure hitherto employed for the computation of the energy of multi-electron atoms (Sect. 2.1.3). This precedent had laid out the analogical correspondence between real and virtual particles which subsequently informed the whole of Nambu's parabola to spontaneous symmetry breaking. Nambu's idea of considering "the meson field as something like the shielded Coulomb field" is partly inspired by the concept of quasiparticle in plasmas introduced by Bohm, which in turn had been shaped after the canonical transformation used in QED (Sect. 2.2.2). Arising from the intersection between the second and the third front, the BCS theory—that Nambu would reformulate in a field-theoretic fashion—was the microscopic quantum theory of a metal that underwent a phase transition, a phenomenon, the latter, which had been traditionally approached with the tools of thermodynamics and statistical mechanics.

When regarded in the context of these two-way isolated exchanges, spontaneous symmetry breaking thus appears as a synthesis that brought the three fronts, with their respective levels of physical description, languages and techniques to significantly bear on each other. A fascinating aspect of this discovery, one that is precisely evidenced by our longue-duree tracking of the epistemic contributions from each front, is the dependence, in fact an interdependence in time, between them on both the formal and conceptual side. That interdependence is what suggests to us the images of an *accretion* and complex *interlocking* to metaphorically render Nambu's path to spontaneous symmetry breaking. On the one hand, quantum field theory provided him with a specific physical picture and the formal tools, in particular the Feynman-Dyson renormalisation scheme and the theorems about gauge invariance, to advantageously treat many-body systems. At a later stage, upon their application to superconductivity, such tools turned out necessary for understanding the spectrum of manifestations that occur in a phase transition, and how these are logically connected. On the other hand, without its combination with Hartree-Fock's approximation scheme, the framework of quantum field theory would not have been applicable in the first place to many-body systems, including those—plasma and superconductors—that would later bear relevance for particle physics itself. Put in more general terms, both the more fundamentally-minded particle physics and the "applied" (or emergent) many-body physics—in particular the concept of quasiparticle invented therein and the mean-field approximation techniques—were necessary to arrive at the comprehension of a crucial manifestation of phase transitions, a comprehension which in turn was instrumental, if not necessary, for revealing the analogous counterpart in elementary particle physics. The emergent many-body phenomena in plasmas and superconductive solid matter that had been formalised by Nambu with the help of fundamental physics, and that had motivated that formalization in the first place, were then recognised as being present in the latter realm too, upon reversing the direction of the analogical implication. The vacuum of the universe was in this way discovered to have aspects in common with the ground state of a superconductor, and the massive elementary fermions and bosons we observe as excitations of that vacuum were found to be substantially analogous to the individual quasiparticle excitations and collective excitations of quasiparticle pairs, respectively.

This epistemic interdependence and heuristic "circularity" of the path to spontaneous symmetry breaking was, albeit certainly an *unexpected* outcome, the result of a precise *programmatic* focus of Nambu. This focus was on dynamical relations between the "atomic" components—regardless of how "elementary" they were, their specific nature and size—and on the search for analogies between such relations. A search that was informed by an awareness of the deepest function of analogical reasoning as a tool that "instructs" the exploration of the unknown in the double sense of giving us instructions and teaching us something of the not-yet-known target. So Nambu will comment about this aspect:

> If phenomena of completely different scales had nothing in common, we would have great difficulty exploring them, even if in the end we might be able to find the answer by sheer observation of the facts. But we can explore the unknown world only with the equipment we already have.

This attitude was essentially what permitted the transfer of knowledge resources—the concepts and mathematical techniques, which Nambu calls the "equipment" in the quote here below—across levels (Nambu 2008). Such an attitude, partly inherited from Tomonaga (see Sect. 5.2.1), demanded a commitment to general concepts and mathematical frameworks more than to a specific level of reality and the related phenomenology, and enabled Nambu to isolate the same (and thus unifying) behaviours at different scales—or anyway between the concepts and techniques we use to rationalise them. In particular, this attitude revealed how the complex behaviour discovered in a complex system can explain some as yet unknown properties of a fundamental and "simpler" system and become the paradigmatic model for it. This "short circuit" between levels where the stream of knowledge flows from more complex to supposedly simpler systems would certainly challenge the reductionist, and might surprise the emergentist too.[2] We witness to the blurring of the primacy of the "fundamental" physics over the "applied" physics—respectively the study of the behaviour

---

[2] The two following aspects of spontaneous symmetry breaking take by surprise both the reductionist and emergence regulative ideals in physics' practice, at least in their strong (and naive) versions. The reductionist, which seeks to explain the behaviour of a (complex) physical system in terms of simpler "atomic" constituents and the interactions between them, would hardly expect a collective many-body phenomenon to be present at the level of the most elementary components, and have any heuristic primacy or relevance for it. The emergentist would expect the emergent phenomena to be prerogative of higher levels and not to be present also in the elementary components.

of the "atomic" components as much as possible in themselves, and that of the phenomena exclusively resulting from compounding several of them—and the breaking of the dichotomous division between the couples individual-simple behaviour and collective-complex behaviour. We likewise are before an instance in which what is discovered is a structural homogeneity between phenomena at distinct levels of physical reality, which manifests itself, in particular, in the presence of a behaviour typical of "mesoscopic" many-body systems—an emergent behaviour—at the level of the elementary particles and their interaction with the vacuum of the universe. It is at least suggestive to see in this instance a reverberation of the old principle of the analogy between microcosm and macrocosm posed by hermetic tradition, and at the same time an anticipation of the idea of universality. If we take up Howard Georgi's idea that "there seems to be interesting physics at all scales", then the discovery of spontaneous symmetry breaking showed how any one such physics can be used to illuminate the physics at another scale, irrespective of how "fundamental" or "derived" the behaviours at these scales are.[3]

In the following section, we reflect on some aspects of the underlying physics that made the "conceptual short circuit" possible; and how this determined an evolution of the concepts of particle and vacuum of quantum field theory.

5.1.1 Few to Many: Spontaneous Symmetry Breaking and the Evolution of the Ideas of Particle and Vacuum

We would like to elucidate in what sense spontaneous symmetry breaking and the analogy superconductor-to-vacuum represented a step in the evolution of the concepts of particle and vacuum with respect to the relativistic quantum framework established with QED, which in turn had represented a significant shift with respect to the framework of non-relativistic quantum mechanics. To properly do so, it is necessary to relate spontaneous symmetry breaking to these two frameworks.

While overcoming in several central aspects (technical and interpretive) the older classical mechanics, the non-relativistic quantum theories developed in the mid-20s by Heisenberg and Schrödinger had remained nonetheless "classical" in one essential aspect which concerns the underlying picture of the elementary bodies. This was the assumption that these bodies with their well-defined intrinsic and measurable properties existed and were conceivable *independently* of and "before" interactions. This was in turn based on the assumption that the latter are weak enough to allow us to assume that whatever interaction did not drastically change the nature of the bodies, that is, transform or determine their core properties. On the mathematical side, this was expressed by the fact that the particles were written in the equation governing the physical system and their properties explicitly appeared therein as parameters.

---

[3] The full quote is: "One of the most astonishing things about the world in which we live is that there seems to be interesting physics at all scales." (Georgi 1993).

Despite being dematerialised— i.e., no longer moving along trajectories and existing as the material objects of classical mechanics—electrons, say, in quantum mechanics existed as free before being subject to the interactions which determined their "course" and were in that sense not treated differently than classical objects, such as planets.

If quantum theory was to be also a relativistic theory, however, a surpassing of the residual classical ontology implicit in the ideas of particle and wave function was needed. In a picture that was both quantum *and* relativistic, particles could transform into other particles but also excite an arbitrary number of virtual particles from the vacuum. A particle propagating between any two points in empty space, i.e., even in the absence of any other particles around to scatter with, undergoes a series of interactions with the virtual particles it itself excites on its course and which consequently co-determine the properties of the particle. In contrast to "classical" objects, the free particles actually observed, according to quantum field theory, were then no longer the bare particles that were a priori written in the dynamical equation but were dressed and modified by the interactions with the vacuum medium. In fact, free particles did not strictly exist as well-defined particles aside from such interactions.

The meaning and purpose of the technical and interpretative expedients of the interaction picture and of renormalization—successfully applied to quantum electrodynamics in the 1940s—was to account for these particle-modifying electromagnetic interactions, implicit in the interaction term, by calculating them, subtracting them from the interaction term and including them in the so-called self-energy of the particle. But since it was not possible to solve for it exactly, the self-energy had to be computed using an approximation. In the absence of solid alternatives, this was the perturbative approximation. While suitable for the calculation of the interactions in quantum electrodynamics, this approximation had, however, two limitations that prevented a more general solution from being seen in other cases. The novelty brought about by Nambu's work essentially derived from replacing the perturbative approximation with a non-perturbative one and embedding the latter in the existing framework of quantum field theory.

The first limitation inherent in the perturbation method, well known and evident, was that it could only properly deal with few-body problems. This was the case since, in order for this method to be used, it is assumed that the first terms of the series, those that quantify the interaction between a few bodies, are the leading terms. For this reason, the perturbative method could not correctly render all the very significant collective effects that derive from the interaction between many components in many-body systems. The second limiting implication, realised lucidly only after the discovery of symmetry breaking, is that perturbative expansion cannot generate outcomes that do not preserve the symmetry of the dynamical equation, regardless of the number of terms of the series that are summed—i.e., irrespective of how fine the approximation is. This happens because each term has the symmetry of the previous one, and therefore ultimately of the non-perturbed term; and the sum of symmetric terms cannot generate properties that are not already in the non-perturbed term. From this followed the crucial consequence that studying the effects on the self-energy associated with a many-body interaction in the perturbative approximation could not give rise to properties that were not already explicitly written in the dynamical equation. In simpler terms, the perturbative computation of the self-energy due to whatever interaction can only *modify* the intrinsic properties of the bare particles that a priori appear in it and cannot account for the *emergence* of the purely dynamical properties. The latter set includes the collective properties like the photon mass in plasmas, the energy gap in superconductors, and the mass of nucleons/fermions in the field theory vacuum. Insofar as they emerge entirely from the interactions and correlations between many particles, these properties do not reside in any of the elementary components, and in some cases, in what appeared as a paradox, they violate the symmetry of the dynamical equation from which they emerge.

Nambu's decisive contribution came into play in this context, to overcome these two limits. In the wake of Tomonaga's research on non-perturbative methods for the study of nuclear systems and in that of Bohm's and Pines' study of collective phenomena in plasmas, Nambu's formalism was set up so as to be a very close non-perturbative analogue of the Feynman-Dyson formulation of quantum electrodynamics. As we have detailed in Sect. 3.2.2, this relied on introducing a (non-perturbative) so-called self-consistent self-energy in lieu of the standard (perturbative) self-energy. This technical device enabled Nambu (i) to reformulate fermionic and bosonic quasiparticles as dressed particles, including therein the collective effects arising from their interaction with the many *real* bodies of a solid; and later on (ii) to apply the same concepts in the field of particle theory[4] and draw out the consequences of such non-perturbative interactions on the properties/symmetries of the elementary particles and the vacuum generating them. One such consequence was the existence of quasiparticle-like solutions which broke the original symmetry of the dynamical equation in those cases where the perturbative method would yield only trivial and symmetric solutions. In this way, Nambu's non-perturbative field theoretical framework demonstrated, for instance, that bare fermions without mass can acquire their mass from entirely dynamical processes. More generally, it revealed the existence of quasiparticles, some properties of which are of purely dynamical-relational origin, that is, they emerge "out of nowhere", entirely from interactions with the vacuum.

Here resides an important implication of spontaneous symmetry breaking. In the parabola we have sketched from the classical particle up to this point, the latter represents a further de-materialisation or de-ontologisation of the physical picture and concept of particle; a progressive erosion, so to speak, of what were formerly thought of as *intrinsic* properties—and consequently appeared as a priori parameters in the free part of the dynamical equation. This de-materialisation consisted in shifting further the boundary between what is intrinsic to the particle and what is a consequence of the interaction with the environment, showing how properties initially considered intrinsic are instead of dynamical-relational origin. A property such as mass, considered completely inherent to matter and to the concept of particle in classical and

---

[4] As we have seen in Sect. 4.2.3, Nambu and Jona-Lasinio applied this technique to the non-linear spinor model Hamiltonian proposed by Heisenberg, where the four-fermion self-interaction term played the role of the electron-electron interaction term in the superconductor.

non-relativistic quantum mechanics, became partially dynamical with perturbative quantum electrodynamics, and then completely extrinsic in non-perturbative quantum field theory. In this sense we should consider the suggestive idea that the work of Nambu, culminating in the model by him and Jona-Lasinio, marked a step in the evolution of the concept of particle in its relation to the vacuum.

It is important to note that this shift occurred while remaining within the framework of quantum field theory, "simply" changing the approximation method in order to extract from the dynamical equations therein the many-body solutions and not only the few-body ones, which are more restrictive than the former. In this sense, the approximation method adopted by Nambu, when applied in the context of particle physics with the Nambu-Jona-Lasinio model, explicated the many-body character that was potentially implied by the quantum relativistic framework, but had hitherto remained implicit. In doing so, an approximation method like Hartree-Fock that had just a numerical function in its original context of use—multi-electron atoms in non-relativistic quantum mechanics—when employed in the framework of quantum field theory instead revealed the "content" of a given dynamical equation (what particles it represents and with which properties) and entwined thereby with the physical interpretation of the equation.[5] In transitioning from the classical to the field theoretical framework, approximation methods change their role: from bridging an equation to one of the possible numerical predictions, to bridging it to one of the multiple descriptions that it implicitly contains and *allows*.

The sense of Nambu's circular intellectual journey, a venture into solid-state physics with the resources of particle physics and then back to the latter with a fecund booty, was that of illuminating particle physics from the outside, uncovering a possibility that was implicit in the folds of quantum field theory. It was, as Nambu and Jona-Lasinio put it in the conclusion of their joint paper, "a new possibility [...] for field theory to be richer and more complex than has been hitherto thought" (Nambu and Jona-Lasinio 1961).

5.1.2 The Heuristic Virtues of a "Scale" Model of the Universe

Spontaneous symmetry breaking shows that when the symmetry constraint is not imposed on the ground state of a dynamical equation, under certain dynamical conditions—an attractive interaction—there are, alongside the symmetrical one, an infinite number of states that are not symmetric, all of them having the same energy. Since this energy is seen to be lower than that of the symmetric state, the actual

---

[5] Elsewhere we have examined the prominent role of approximation methods in explicating what we call "the representational content" of a given model equation; and remarked how the mutated relationship between the latter and the approximation method warrants a revision of the view according to which approximation methods are simply numerical tools—entirely divorced from questions of physical interpretation nor expected to yield fundamental insights (Olano et al. 2022).

ground state must be one among the infinity of such asymmetric states. Thus it is shown that at a sufficiently low temperature, for example, the ferromagnetic or superconducting state—states that respectively violate the rotational and gauge symmetry (the conservation of angular momentum and electric charge)—become the stable state of a solid while the normal state becomes the unstable state; and that stable state is characterized by a specific direction of magnetization or, in the case of the superconductor, a fixed macroscopic phase of the wave function. In the laboratory, with a table-top experiment or a cryostat, this phenomenon of "spontaneous polarization" happens before our eyes and is measurable. Going across the critical point between one phase and the other, one passes from the normal state to the asymmetric state, and in the absence of a strong external perturbation one can in principle inspect a multiplicity of these physically distinct asymmetric states.

In the analogical extension of this idea to the entire universe that Nambu and Jona-Lasinio operated in their model of elementary particles, the "true" state of the universe (the "world") in which we live must be one in the infinity of such asymmetrical "worlds" equally probable and indistinguishable from each other. Adopting the explanation that some manifestations of the universe, like the masses of elementary particles and the corresponding Nambu-Goldstone boson (the Higgs boson), are consequences of spontaneous symmetry breaking, as the gap and magnetization are in their respective solids, therefore implies the existence of both an unstable world in which particles have no mass and an infinite series of other worlds which are all physically distinct from ours and "outside of each other", as Nambu and Jona-Lasinio write. This metaphysical flavour comes from the fact that "no interaction or measurement [...] can bridge 'these worlds' in finite steps", unlike what happens in solids for which, by changing the external conditions "from the outside", such "worlds" can be inspected and we can have experimental access to them.[6]

While spontaneous symmetry breaking is today as much a key to interpreting phenomena in high-energy physics as it is in solid-state physics, this cognitive asymmetry remains and is arguably reflected in the heuristic path that Nambu followed. It might, then, not have been a chance that spontaneous symmetry breaking was after all discovered in a solid-state domain before it was recognised in the domain of particle physics, and it was the former that served as the blueprint for the latter, rather than the other way around. This heuristic primacy resides in the relative "experimentability" and manipulability of solid-state physical systems. The compact superconducting samples placed in our laboratory cryostats can serve as a concrete "scale" model of the universe (a microcosm), but the reverse would hardly happen. While we sit outside of such systems and can thus manipulate them by varying the external control parameters, we cannot isolate the elementary particles from the vacuum of the universe or manipulate the control parameter of the phase transition, and the latter instances will be inevitably more abstract than the former. We are immersed and

---

[6] With some fantasy, we can compare this type of explanation to the one that imaginary inhabitants of a two-dimensional surface would give of three-dimensional objects, being able to deduce the existence of the latter despite not being able to observe them by observing how in their world a one-dimensional object can be the result of a two-dimensional object. The paragon is inspired by the inhabitants of Edwin Abbot's two-dimensional Flatland.

locked into the low-symmetry phase of the universe, with only indirect manifestations and speculative tools to grant access to the other phase(s). In the final leap to spontaneous symmetry breaking, the analogical extrapolation of the conceptual scheme of superconductivity was cleverly used as one such tool to look, as it were, at the universe from the outside and abstract ourselves from it.

As we stressed above, however, the epistemic flow was not one-directional and the dependence was mutual. As much as the "closer" domain of solid-state physics offered concrete and controllable instances of spontaneous symmetry breaking in the form of phase transitions, the more "distant" and abstract domain of particle physics had provided to it the powerful and effective formal tools to understand them. It is in this sense quite clear why a ubiquitous phenomenon like spontaneous symmetry breaking in solids had not been discovered before. We are before a case of interdependence of knowledge resources.

5.2   Methodologies and Worldviews on the Japanese Shores

As in the nature of any heuristic speculative activity, some crucial steps and turns of the parabola to the discovery of spontaneous symmetry breaking we have traced in this book hinged on broader underlying philosophical and cultural stances. While it is often hard to measure and lay claim to such influences owing to the subtlety in which they act and their "diffused" nature, we feel that two of these emerge here with sufficient distinctness and clarity. The first is the influence of the school of Tomonaga, and the second is the influence of Yukawa and Sakata's school. Distinct in their methodologies, heuristic approaches, and world-views, these schools in fact manifest broader cultural orientations and contextual factors that are peculiar to the Japanese situation. Ultimately, these schools are living examples of how the grafting of the (broadly intended) Western discipline of modern physics on to the *local* context of Japan—in what was perhaps the last big wave of globalisation of scientific knowledge—contributed to transform that discipline, thereby signalling an active appropriation rather than a passive colonisation. As it was consummated on the ground of Western science, this marriage meant espousing in the local context the rules of the "game" of Western science and the philosophical premises underlying it: a certain logic and methodologies, an ideal of objectivity and a clear distinction between subject and object, the categories of substance and essence and the very possibility of (scientific) "stable" knowledge of the essences condensed in the Platonic-Aristotelian doctrine of *epistéme tes aletheis* (literally, the knowing of the truth). On the other hand, this adoption did not prevent an interplay with the traditional underlying way of thinking founded on different—and sometimes perpendicular—stances and attitudes. This interplay is the one we see unfolding at various orders in the following two sections.

5.2.1 The Influence of Tomonaga and the Local Socio-Economic Context of Japan

As we have already tracked in the course of our story, three interdependent motifs are relevant in drawing the line of continuity between Tomonaga's tradition and Nambu's work. One is the use of knowledge resources (concepts and mathematical techniques) from atomic physics and solid state physics to approach problems in nuclear and particle physics. The second motif is the systematic use of the real↔virtual analogical correspondence to carry out, sometimes with formal intent, sometimes as an intuitive lever, the transfer of such concepts and techniques. As we have detailed in Chap. 2, these two heuristic strategies were systematically employed by Tomonaga. Among other applications, he used them to construct a field-theory analogue of the Hartree-Fock method for the calculation of the self-energy of a nucleon; to set up renormalisation in quantum electrodynamics having in mind what he had seen as an analogous procedure in the calculation of energy spectra of multi-electron atoms; and to apply the approximation method that Felix Bloch had used for charged particles in solids to complex (strongly-interacting) nuclei. The preconditions for the use of such strategies seem to be quite a mathematical predisposition as well as an interest in the "crafting" of mathematical frameworks and methods of approximation. These characteristics are the third distinctive element of Tomonaga's influence.[7]

If these three elements had a significant influence on Nambu and found their most fruitful expression in his work leading to spontaneous symmetry breaking, they can also in various degrees be recognised in other physicists who, like Nambu, had been part of Tomonaga's study group. In fact, rather than an exclusive prerogative of this school, these elements signal a broader influence of the socio-economic context of Japan at the time, of which the school of Tomonaga is the clearest expression. The peculiarity of the context had to do with the material conditions, the marginalisation and relative isolation of pre- and post-war Japan as compared to the Western contexts. Due to the scarcity of economic resources, academic posts and career prospective, in contrast with their homologues in the United States and Europe, the trained Japanese physicists had on average much less certainty about the academic or industrial branch they would end up working in. Besides, due to the relative absence of well-defined curricula, during their studies they were inevitably exposed to a variety of problems and techniques from low-energy physics to high-energy physics. For this reason, they were compelled to be as versatile as possible and to make the best out of what they had at their disposal.

This contextual difference in Japan was concretely reflected in a segmentation of the sub-disciplines of physics in which the boundaries were much more fluid as compared to those witnessed by Western physicists. This contrast was especially sharp for the generation that had begun to work in the post-war period. Both in the United States and in Europe, under the wartime and post-war investments and the employment of a massive scientific workforce, physics in the 1940s was already

---

[7] The epithet of "craftsman" to qualify Tomonaga is used by Nambu himself (Nambu 2008).

well on its way to being fragmented into various low-energy and high energy sub-disciplines, with the division being drawn according to the energy scale and size of the physical systems investigated and the type of interaction (many-body the former, few-body the latter) characterising them. Conversely, in Japan—and in particular in a few places and institutions—the distinctions between low-energy and high energy physics could hardly be enforced so sharply, and physicists continued to be routinely exposed to both in their educational and professional life (Low et al. 1999; Low 2005). This resulted in a type of segmentation of the realms of physics which is alternative, or "transversal", to the one set by academic disciplines over the course of the 20th century. Solid-state systems, molecular, nuclear, and sub-nuclear systems all belong to distinct disciplines when regarded from the point of view of size and energy scales, but partially overlap when regarded from the point of view of the underlying dynamical processes—which are largely independent of the scale and nature of unit components—and mathematical framework used to approach them. This underlying peculiar element of the Japanese context, and the specific inheritance of Tomonaga's thought style and related inclinations, forged a substantial part of Nambu's characteristic mode of enquiry.

5.2.2   On Yukawa and Sakata's World-view and Its Sway

If we have discussed the more transparent handover between Tomonaga and Nambu quite extensively throughout the text (Chaps. 2 and 3 mainly) and recalled it above, there is also another influence, complementary to the former, which is no less important, but which we have not discussed so far. As it blends in with the broader Japanese cultural tradition, this influence is more subtle and difficult to tease out. It has not so much to do with the use of analogy, but with (i) the conception/image of the quantum vacuum, and (ii) the more general heuristic method and the underlying philosophy of nature, both of which are greatly conditioned in Nambu by the thought of Hideki Yukawa and Shoichi Sakata. Let us first give a sketch of this thought, underlining the rhetoric and the ideological charge that accompanied it and through which it got hold of the young generation of physicists, including Nambu.

Sakata and Taketani Heuristic Method and Philosophy of Nature

With these words Shoichi Sakata, close to his death, commences his address to the Japanese National Radio channel on July 30, 1969[8]:

> Physics of atomic nuclei and elementary particles, which is my speciality, is entirely a new field of physics started more or less at the time when I was an undergraduate university student [early 1930s]. Since one of our greatest concerns has been how to overcome the

---

[8] The speech was published in Japanese on the March 1971 issue of *Kagaku* (Science) and then in English in (Sakata, 1971a).

old, it may seem that we do not have much to do with the classics. But, creating new things through overcoming the old is the central problem in the method of science or methodology. In this sense, the classics are very important for us. As one of my classics, I want to quote Engels' Dialektik der Natur (dialectics of nature), which has been continuously sending invaluable light into my studies of about forty years as a precious stone. Today, I would like to talk about how I encountered Dialectics of Nature and what influences it gave to my studies in physics.

What might seem to the unacquainted reader an unusual, if not preposterous, beginning, or just the political wanderings of an old progressist physicist-activist, does in fact signal an important element of the development of Japanese physics, which suffused through a generation of Japanese physicists, and significantly influenced a few of Nambu's turning points.

The interest in the Marxian view of nature and science was indeed raised in Sakata through Friedrich Engels' "Dialectics of Nature"[9] and by the physicist and philosopher Mitsuo Taketani, whom Sakata befriended in the early 1930s (Sakata 1971a). A peer of Sakata and collaborator of Hideky Yukawa in the years of the elaboration of the meson theory, Taketani issued what would be considered his contribution to physics in 1936 in the articles "Dialectics of Nature-on Quantum Mechanics" and "Doctrine of the Three Stages of Scientific Development". There, Taketani put forward the thesis according to which both classical and quantum mechanics had developed by going through three stages. Inspired by the Hegelian triad, these were the stages of Phenomenon, Substance, and Essence. The first stage consisted in finding phenomenological relations which arrange the experimentally observed regularities in a set of formulae. This is followed by the search for the hypothetical material substances at the origin of the given phenomenology, or the investigation of the "structure of the object" (Sakata), in what is called the substantialistic stage. In the third and last, the so-called essentialistic stage, the dynamical laws that govern such substances and explain the phenomena from a more universal point of view are formulated. At every stage of the process, one description surpassed the other in view of the synthesis between the phenomenological and substantialistic perspectives to be found in the essentialistic stage. Once obtained, this synthesis awaited a new phenomenon or a contradiction to open the underlying layer, and so on without an absolute and foreseeable end.

To this methodology laid out by Taketani, a specific ontological picture of the natural world corresponded. Consistent with Engels' dialectics of nature—to which Taketani and Sakata subscribed—the world was conceived as being constituted by an infinite series of strata, each with its own laws, states and particles. Particles that are thought to be elementary are discovered to be composite when the layer below is accessed, and so on, with no ultimately authentic elementary component. An analysis of the historical development of science through the "stereoscopic view of the materialistic dialectics" revealed such a structure and would continue to do so in the future. In light of this, as Taketani argued, the heuristic methodology to adopt in order to most swiftly and flawlessly progress through these levels had thereby to mirror this

---

[9] This text was translated in Japanese as early as 1929, just a few years after its (posthumous) publication in German.

endless stratification of the physical reality and follow at each level the three stages from phenomenological to the essentialist, passing through the *substantialistic* stage. As we will see in a moment, the presence of the latter is the really original element conveyed by this methodology and the one which reflected a characteristic attitude of Japanese thought.

Sakata viewed in the methodology proposed by Taketani "a compass" and regulative principle when searching for yet unknown laws of physics. Differently than that of the positivists and the realists, according to Sakata, this methodology defined a clear heuristics, which he spread to others and used himself as a guiding principle for his research in theoretical particle physics. Once acquired full consciousness of this methodology and world picture, that was Sakata's underlying credence, and seeing how the lack of awareness of them had caused unfounded surprise, crisis, and struggle among the Western scientists, it would be erroneous to be guided by anything other than that. So he had been repeating with determination for several years[10]:

> [...] Now that the great fruits of modern science have proved the validity of "dialectics of nature" and therefore revealed that the cognition of nature is made through the dialectic processes, we must intentionally apply the dialectics of nature as a compass which shows the way of our research.

The "new view of nature and a correct methodology" was ardently promoted by Sakata and Taketani among Japanese physicists in opposition to both the empiricist and realist alternatives assumed in Europe and incarnated respectively by the figures of Ernst Mach and Max Planck (in a debate that had left a strong impression on Japanese scientists). Against both the "old and fossilised" realist view assuming the existence of ultimate indivisible material particles and the empiricist view of such components as a mere working hypothesis, Sakata and Taketani's new view and methodology argued in favour of the elementary particles observed as being real, but only one stage to something else in the cognition of nature. At bottom, the accusation was directed to the difficulty of "Western thinking" to come to terms with a physical reality with infinite strata, in which every stratum is real and yet never the ultimate, so that every postulated entity is provisional.

Needless to say, the world picture promoted by Taketani and Sakata suffered the limits of the ideological and dogmatic interpretation of physical and historical reality. It featured various ambiguities and was questionable in its application to the retrospective analysis of the development of physics as well as in its meta-inductive significance. Despite and aside from all that, this methodology and world picture did provide Japanese physicists—including Nambu—with an original angle on some concepts and questions that they had inherited from Western physics, and resulted in a different conception of, among others, the vacuum, the Dirac sea, elementary and composite particles. The reason for these original conceptions and this angle, however, had less to do with the Hegelian-Marxian-Engelian ideology in itself—as Sakata and Taketani stated—than with its resonance and combination with pre-existing traits

---

[10] The following is an extract from the October 1947 issue of the journal *Cho-ryu* (Current). It has been translated into English in (Sakata, 1971b). Other essays by Sakata and Taketani on the same topic can be found in the same volume.

of Japanese thinking. That ideology happened to incorporate and espouse an attitude and conception of nature (and of the cognition of it) that Japan had inherited from the Chinese and Buddhist thought; and as such it at most helped to give a frame and pride to two traits of this attitude and conception of nature in the context of modern physics.[11] These two deeper cultural traits of Japanese thought were quite foreign to the Western speculative tradition that had largely informed modern science, and they played a role in Nambu's original conceptions and attitudes.

One of these traits is, roughly put, the tendency toward conceptual representations that are concrete rather than abstract. In other words, the preference for conceptual expressions which concretely represent thought, and conversely a general refusal of abstract categories. Borrowing the words of the illustrious scholar Marcel Granet in a penetrating study of Chinese thought, the verbal or written expressions in traditional Chinese thought aim "at representing thought" in a way that "this concrete representation imposes the feeling that expressing or rather depicting is not simply evoking, but arousing, realizing" (Granet 1934). These concrete representations in the Chinese language invariably "involve a series of other visions all equally concrete". On the basis of the importance given to the emblems, Granet highlights how "Chinese thought refused to distinguish the logical from the real" and "did not want to consider Numbers, Space and Time as abstract entities", nor was it considered "useful to create abstract categories such as our categories of Genus, Substance and Force."

The other trait that we focus on, which in fact is not unrelated to the first one, is the tendency characteristic of Buddhist epistemological doctrine to dismiss absolute and ultimate entities as artificial constructs and non-existent in reality. This arises as a consequence of the belief in the interdependent character of every entity (whether material or spiritual), from which stems their impermanence and the impossibility of absolute and metaphysical essence (or unchanging self) (Pasqualotto 1992). If

---

[11] That the Hegelian (and then Marxian) view had a kinship with Buddhist and Chinese thought is not coincidental. In the first place, differently than what is commonly believed, Hegel performed a profound study of Eastern thought which influenced his philosophy (see, e.g., De Pretto (2011)). Secondly, in Japan, philosophy (*tetsugaku*) for how we intend it in the Western tradition—as constructed by the largely influential exponents of the Kyoto School of Philosophy—largely moved from the adoption of the idealistic doctrines and its fusion with long pre-existing elements of the Buddhist thought. This was arguably promoted by the resonance of some basic conceptions of Idealism (adopted and co-opted then in the dialectical materialism) with those of the traditional Chinese wisdom and Buddhist traditions. These are essentially the co-dependent generation of all the worldly objects and phenomena, from which descends their intrinsic interconnected and interdependent character, ceaseless transformation, impermanence (Pali *anicca*) and absence of an absolute and metaphysical existence (Pali *anatta*)—that is, an existence unthinkable outside of the relation (Pasqualotto 1992). Especially in Buddhist thought, these are held as the only fixed elements of the universe (and should not be intended as abstract laws, but as laws of eminently practical order). With regard to the kinship between German idealism and traditional Buddhist thought, the philosopher Hiroshi Nagai writes (1952): "Why was German idealism easier to understand than English empiricism? Was it not due to the historical tradition of our thought which had been based on a certain metaphysical speculation of Buddhism?". In general, we might say, using an expression from Soshichi Uchii, that philosophy in Japan resulted from the "understanding and misunderstanding" of the "new discipline of [Western] philosophy" (Uchii 2002).

entities are secondary and variable, primary and invariant in this epistemology is the dynamical principle out of which entities and phenomena co-dependently emerge and transform. According to this principle of interdependent co-arising/co-causation (Pali *paticcasamuppada*), entities and their phenomenal properties are thus considered as consequences of interaction processes and as having no independent and stable existence. The attention of traditional Buddhist thought is focused on such interdependence.

Nambu's Synthesis

These two elements of the cultural tradition, suffused in Japanese thought and Zen tradition, in their own ways are reflected most evidently in Yukawa's heuristics first, Sakata and Taketani's methodology and world picture, and then in Nambu's thinking. The first element, in particular, found its clearest expression in what Sakata and Taketani rationalised as the "substantialistic" stage. This stage consists, as Nambu would later put it, in "trying to find an explanation for the origin of a given phenomenon in terms of some concrete objects" as a way to reach the precise and detailed mathematical framework (which in Sakata and Taketani's three-stage theory was associated with the "essentialistic" stage). The second element consists of working with the assumption that there is an infinity of such different entities, some for every layer, rather than with the assumption of the existence of an absolute and ultimate substance. The entities assumed at a given moment are thus, in this view, constitutionally provisional and instances of an endless dialectical process which will reveal other, and different, ones at the underlying level. As Nambu vividly conveyed, this accorded a sense of historicity and mutability to scientific truth (Nambu 2010):

> [Uncertain about a daring conceptual move] I recalled an edict by S. Sakata. He said that truth is not immutable but will evolve. What we think now as truth is not absolute; it may be modified later as we learn more. Reassured by Sakata's words, I decided to go ahead.

> From now on our task will be to check the Standard Model in more detail. If disagreements are found, it will then signal the beginning of the next cycle, as is expected or hoped for by everybody.

That mixture of elements betrayed the quite peculiar attitude that some Japanese physicists had towards the entities of physics and provided them with different and alternative conceptual schemes to work with. This granted a certain power and freedom to them when it came to positing new entities as compared to the West where, under the sway of Occam's razor and the "horror" of the multiplication of entities, there was quite a reluctance to do so. Certainly the renowned and most revolutionary example of that was the postulation of Yukawa's π-meson in the mid-30s, but that would be followed by several other instances—the C-meson and the μ-meson (the modern muon) of Sakata were two such instances.[12]

---

[12] Nambu will recollect the first encounter he had, when he was 21, with Sakata's way of doing physics in connection with the postulation of the muon (Ashrafi 2004): "[...] just a few months before I was drafted into the Army, the summer of 1942, Tomonaga read a letter from Sakata saying

We can regard the different course taken in Japan by the idea of the Dirac sea—the picture of the vacuum as a Fermi gas—and the further elaboration by Nambu as being the product of this same attitude towards physical entities. The contrived construct of Dirac, that had been surpassed by Western physicists had instead survived there. Hara and Shimazu, from the school of Sakata, still in 1952 write, for example: "Let us consider [...] the interaction of the electromagnetic field with the electron-field. The vacuum is defined according to Dirac's positron theory as a state in which all negative energy states are occupied. This state is however nothing but a sort of Fermi gas at the absolute zero of temperature" (Shimazu and Hara 1952). Nambu himself, too, in two recollections many years later, explains what was his position towards that picture (Brown et al. 1992; Brink et al. 2016)[13]:

> Under the influence of the Yukawa-Sakata school I had always tended to seek physical substance under formal mathematical statements. Thus Dirac's assumption of the filling of the negative energy states in the vacuum [...] was to me not a mathematical trick but a reality. I am always tempted to think of concrete objects behind mathematical expressions. Thus I was fond of Dirac's interpretation of the vacuum as a sea of negative energy states.

As we have seen, Nambu adopted Dirac's picture as the starting point which led him, through subsequent processing, to finally identify the vacuum of the universe not with the simple Fermi gas he had started with, but with the ground state of a superconducting solid—i.e., where the negative energy states are populated not with standard fermions but with analogues of Bogolyubov quasiparticles. On the basis of what we have said above about the tendency to prefer concrete representations to utterly abstract constructs, it is hard not to read in Nambu's constant thinking of the vacuum *as* a solid media that traditional cultural trait, re-emerging in the context of modern science as a general heuristic principle.[14]

Considering that Dirac's hypothesis had been highly problematic from the outset, this way of thinking and in particular the adoption of the hypothesis as heuristic stepping stone—a lever, we might say—were not obvious. They required the *suspension* of precisely those questions about the ontological status of the vacuum electrons/nucleons, their real or fictitious existence, and their observability that had raised objections among Western physicists, objections which were certainly mainstream at the time. They also demanded a degree of interchangeability between the ideas of vacuum and matter, combined with a skillful use of the analogy: the virtual nucleons filling the vacuum and real nucleons were both simultaneously regarded as real matter, when considered under the aspect of the Pauli principle, and as virtual

---

that the cosmic ray mesons and Yukawa's particles are two different things. This is the so-called two-meson theory. And he explained to us what that theory was, and I was very impressed."

[13] It is important to emphasise that such a position is not the consequence of a retrospective revision of his own work. As we have seen in Sect. 3.2.1, uses and appeals to that picture appear first in a 1950 brief communication (Nambu 1950), to then recur throughout the following decade.

[14] As evidenced by Nambu's subsequent thinking, for instance in relation to his early speculations in string theory, this seems to have been a trademark of Nambu's thought style. Assays of it are given by students and collaborators of his in Brink et al. (2016).

matter, when considered under their contribution to the self-energy; correspondingly, the vacuum was considered to act like any other real medium and vice versa the latter was considered as the observable manifestation of such a vacuum—a second, or "apparent" vacuum as Nambu called it.

As we have detailed, this rather light-hearted and playful recovery of Dirac's concrete representation of the vacuum was judged by Nambu not on the level of the dichotomy between physical and unphysical—that is, as being true or false—but considered as a provisional posit apt to look at the many-body forces through the concept of self-energy, and, more generally, to use the hitherto perfected machinery of quantum electrodynamics to arrive at a generalised and neat mathematical framework for solids. With the knowledge gleaned from this operation, it is with the same spirit and lever of concrete representation, only used in the opposite direction, that a decade later Nambu re-mathematised the vacuum of the world, revealing how in a certain sense it really is a superconductor. Borrowing from Archimedes, we can say that heuristic lever was indeed long enough to lift the world.

References

Ashrafi, B. (2004). Interview of Yoichiro Nambu by Babak Ashrafi on 2004 July 16. www.aip.org/history-programs/niels-bohr-library/oral-histories/30538.

Brink, L., Chang, L., Han, M., & Phua, K. (2016). *Memorial Volume For Y. Nambu*. World Scientific Publishing Company.

Brown, L. M., Brout, H. R., Cao, T. Y., Higgs, P. W., Nambu, Y., Hoddeson, L., Brown, L., Riordan, M., & Dresden, M. (1992). Panel session: Spontaneous breaking of symmetry. In *The rise of the standard model: Particle physics in the 1960s and 1970s* (pp. 478–522). Cambridge University Press.

De Pretto, D. (2011). *L'Oriente assoluto: India, Cina e "mondo buddhista" nell'interpretazione di Hegel*. Mimesis. Pensieri d'Oriente. Mimesis.

Georgi, H. (1993). Effective field theory. *Annual Review of Nuclear and Particle Science, 43*(1), 209–252.

Granet, M. (1934). *La pensée chinoise*. La Renaissance du livre: Bibliothèque de synthèse historique.

Low, M. (2005). *Science on the international stage: Hayakawa* (pp. 169–196). New York: Palgrave Macmillan US.

Low, M., Nakayama, S., Yoshioka, H., et al. (1999). *Science, technology and society in contemporary Japan*. Cambridge: Cambridge University Press.

Nambu, Y. (1950). Pi-meson self-energy (in Japanese). *Soryushiron Kenkyu, 2*(2), 170–174.

Nambu, Y. (1995). A 'superconductor' model of elementary particles and its consequences. In *Broken symmetry: Selected papers of Y. Nambu* (pp. 110–126). World Scientific.

Nambu, Y. (2008). The legacies of Yukawa and Tomonaga. *Americal Physical Society Bulletin, 18*(6), 7–11.

Nambu, Y. (2010). Energy gap, mass gap, and spontaneous symmetry breaking. In *BCS: 50 years* (pp. 525–533). World Scientific.

Nambu, Y., & Jona-Lasinio, G. (1961). Dynamical model of elementary particles based on an analogy with superconductivity. i. *Physical Review, 122*(1), 345.

Olano, P. R., Fraser, J., Gaudenzi, R., & Blum, A. (2022). Taking approximation methods seriously: The cases of the Chew and Nambu–Jona-Lasinio models. *Studies in the History and Philosophy of Science, 93*, 82–95.

Pasqualotto, G. (1992). *Estetica del vuoto: Arte e meditazione nelle culture d'Oriente*. Marsilio.

Sakata, S. (1971). My classics—Engels' "Dialektik der Natur". *Progress of Theoretical Physics Supplement, 50,* 1–8.

Sakata, S. (1971). Theoretical physics and dialectics of nature. *Progress of Theoretical Physics Supplement, 50,* 103–119.

Shimazu, H., & Hara, O. (1952). On Yukawa's theory of non-local field. *Progress of Theoretical Physics, 8*(3), 385–386.

Uchii, S. (2002). Is philosophy of science alive in the East? A report from Japan. 40th Anniversary Lecture Series, March 25, 2002, http://philsci-archive.pitt.edu/585/.

Printed in the United States
by Baker & Taylor Publisher Services